# The Roots of Visual Depiction in Art

# The Roots of Visual Depiction in Art:

## *Neuroarchaeology, Neuroscience and Evolution*

By

Derek Hodgson

**Cambridge Scholars** Publishing

The Roots of Visual Depiction in Art:
Neuroarchaeology, Neuroscience and Evolution

By Derek Hodgson

This book first published 2019. The present binding first published 2021.

Cambridge Scholars Publishing

Lady Stephenson Library, Newcastle upon Tyne, NE6 2PA, UK

British Library Cataloguing in Publication Data
A catalogue record for this book is available from the British Library



ISBN (10): 1-5275-6208-5
ISBN (13): 978-1-5275-6208-0

For Greta, Mia Rose and Lucinda

# TABLE OF CONTENTS

# LIST OF FIGURES

# PREFACE

Almost twenty years ago, I began to look at theories that sought to explain the appearance of the relatively naturalistic portrayals of animals during the Upper Palaeolithic in Europe. Despite the fact that the depiction of animals is a quintessentially visual phenomenon, to my surprise, I found almost no reference to recent findings from the neuroscience of vision. Except for a few papers that explored Palaeolithic art from the perspective of perceptual psychology, and probably because, back then, neuroscience was still in its infancy, scant regard was given to the potential insights to be gained from the ever more powerful brain scan techniques. I set out to change this by illustrating how neuroscience could provide novel insights into the reason ancient hunter-gatherers first began to create images of animals. But this could only be achieved by taking into account the deep evolutionary precursors influencing brain function as well as the ethological and environmental context that influenced behaviour. Since those early days, the pace of change in neuroscience has opened up fresh fields of analysis, which are uniquely placed to provide insights into Palaeolithic art. The application of those developments to archaeology became even more imperative given the exciting new discoveries of Palaeolithic art over the past two decades. One of the main barriers to acceptance came, at first, from archaeologists themselves who, understandably, preferred not to stray too far from the material record. However, thanks to a growing interest in cognitive evolution amongst both archaeologists and anthropologists, neuroscientific insights gradually came to be accepted to the extent that a sub-discipline was formed referred to as neuroarchaeology. As a result, neuroscientific research began to be applied to archaeology in areas beyond Palaeolithic art. A further hindrance to acceptance was the fact archaeologists and anthropologists tended to favour "symbolic" interpretations, which meant that archaeological finds continued to be considered from a top-down, higher-order perspective according to socio-cultural criteria. This had the effect of ignoring preconscious bottom-up factors, which neuroscience established has a profound influence on conscious behaviour. I therefore set out to assess whether and to what extent particular aspects of Palaeolithic depictions might be amenable to an analysis from the purview of neuroscience. The contents of this book present the culmination of almost two decades of research that illustrates how an approach stemming from recent neuroscientific discoveries can reap rich rewards for understanding the origins of the earliest figurative art.

Apart from the first part of Chapter Four, entitled "The 'expedient' and 'discerning' visual brain", which can be skipped without compromising the essence of the book, the text has been written to appeal to the general reader as much as to those with a specialist interest in the origins of figurate depictions. The more challenging technical concepts have been relegated to the footnotes where those interested can find information regarding underlying principles. The first four chapters provide a grounding for the core of the book, which is to be found in Chapter Five and Six containing the most important insights. The chapters following explore the implications arising therefrom.

# CHAPTER ONE

# INTRODUCTION

> The dominance of depictions of animals in prehistoric art suggests that the advantages of the animal connection may have driven the development of figurative art… (Shipman 2010, p. 524)

## Paradoxes, progress, and definitions

Despite the extraordinary controversy regarding Upper Palaeolithic (UP) art since its discovery in the 19th century, progress towards explaining its provenance has been limited. The lack of progress may be because, except for some very early and contested sculptural artefacts dating to several hundred thousand years ago, fully fledged iconic depictions—in the form of figurines and two-dimensional representations—seem to appear suddenly around 37,000 BP (before present). What is baffling is that many of the first representations were skillfully naturalistic and persisted in virtually the same format for 30,000 years. Having said this, increasing evidence from the archaeological record suggests artistic behavior, which is not in any way figurative, has roots far more ancient than previously surmised. Typical examples consist of ornamental shells and beads as well as ochre used for body colouration, and there is growing evidence for the existence of geometric marks from different parts of the world that predate the depictions of the UP, perhaps dating back 500,000 years. How do these early examples of "art" relate to the much later incarnations? Are they merely isolated occurrences or do they indicate something more fundamental?

A further contentious issue concerns the lack of an accepted definition of art. A problem exacerbated by the fact that, in the west, art has become the preserve of specialist gatekeepers who, based on the prevailing consensus, decide what is to be accepted as such. In fact, today's art is commonly regarded as a luxury associated with increased leisure or greater wealth. Even worse, the skill of making art and a concern for aesthetics are all but derided, making art inscrutable to the wider population. This situation has been muddled further by the fact that, in the modern era, different art practices have become separate areas of concern whereas for early hunter-gatherers such a separation did not exist in that various forms of expression were coextensive. When referring to the art of early hunter-gatherer groups we therefore need to bear in mind that a particular form of artistic activity is invariably linked to other modes of expression as part of a socio-cultural matrix.

The fact that for most of human history art was practised by small scale traditional groups suggests that, in terms of definition, the longer-term needs to be prioritized over the much shorter modern period. In fact, most art was utilized by communities engaged in a hunter-gatherer lifestyle where it was integral to daily life, which is all the more surprising considering the arduous environmental conditions hunter-gatherers had to endure during the Ice Age. The problem of an acceptable definition of art has also been exacerbated by the tendency to ignore how "non-functional" materiality was employed as a means of expression in small pre-literate groups from 42,000 years ago up to the beginning of the historical era; by far the longest period compared to modern western forms of art. In today's terms, art tends to be regarded as non-functional, whereas for pre-modern traditional communities art is taken to be "functional" in the sense that it arguably provided a way of expressing group identity. An approach to understanding art as a behaviourally complex phenomenon, which includes figurative depictions, therefore needs to be based on its origins deriving from at least 42,000 BP up until when this began to change during the early Neolithic period and beyond.

In addition, we need to realise that the art of hunter-gatherers and small scale groups is collective in that all members of a community are involved with various kinds of expression, which are usually experienced co-currently e.g., singing, dancing, visual arts etc. Some activities, however, may occasionally be practised separately by several members within a group. Thus, rather than "art" we would be better served to refer to "arts" when addressing this issue in the present context. A consideration of the role of the arts should therefore be based on when complex multifaceted art first arose around 42,000 years ago. Having said that, we need to take into account the fact that art-like behaviour occurred well before the UP but which was fairly restricted in terms of complexity (Conard, 2008). In order to avoid confusion with the modern usage of the term "art", throughout this book we will often refer to palaeoart, which includes the UP art of Europe, as well as all earlier aspects of rock art and manifestations of artistic behaviour (such as personal ornamentation and geometric motifs) found at many ancient sites globally.

Moreover, since 42,000 years ago—perhaps longer—the arts in one form or another, were practised by all social groups. This is surprising considering the cost in terms of time and effort incurred that could be more profitably directed towards more obvious survival activities. One clue to derivation can be found in the fact that creating and experiencing the arts is inherently pleasurable to the extent that individuals spontaneously engage in such activities without receiving an obvious reward.

A further point needs to be taken into account, aesthetics was not fundamental to hunter-gatherers (Brown and Dissanayake, 2009), which is not to say there was no interest in aesthetics, only that aesthetics remained secondary to the main goal, namely the way material culture helped structure ongoing lifeways (Hodgson and Verpooten, 2014). This is supported by the fact that, although some areas of the brain are dedicated to aesthetic awareness (Ishizu and Zeki, 2011; Brown et al. 2011), no area exists dedicated to a particular art form or the arts in general (Zaidel, 2010).[1]

Understanding the role of art is important because whether it is biological or not is critical in establishing why it prevailed. A number of scholars have taken the biologically adaptive route (see, for example, Dissanayake,1999, 2000; Miller, 2001; Boyd, 2005; Dutton, 2009), though such an approach has been subject to considerable criticism for a number of reasons (see Davies, 2012). An alternative, and potentially more productive, line of enquiry asserts that the arts need to be seen as part of the broader spectrum of material culture that arises from gene-culture co-evolution (Boyd and Richerson, 2005, 2007; Richerson and Boyd, 2001). Accordingly, culture itself is regarded as adaptive but in a special sense—an idea to which we will return throughout this book

One further complication concerns the role of iconic representation, which is often held to be central to the plastic arts. Paradoxically, many small-scale traditional groups up until quite recently did not produce figurative art but relied on elaborate geometric patterns to express their artistic inclinations. Nevertheless, the discovery of figuration represents a crucial boundary in artistic expertise that needs to be contextualized. All the more so, since figurative depiction is intimately associated with art in that it was exploited in many different ways by different groups from around 40,000 years ago onwards. Consequently, a significant portion of this book will be devoted to examining the emergence of iconic depictions as they provide a clue to the first appearance of "complex" artistic behaviour.

The fact today's visual brains are, despite a few caveats, essentially the same as those responsible for the first depictions at the end of the Pleistocene (ending around 12,000 year ago), provides an opportunity to delve into the very minds of our forebears. Stable attributes of this order will allow us to offer some intriguing suggestions as to what art meant to early communities. The first "art" may not only help inform us about those responsible for producing the earliest depictions but also assist in elucidating how early hunter-gatherers related to their world. This is because representational art is intimately associated with how the visual system functions. Indeed, as vision in humans constitutes the dominant sensory modality that operates at an extremely sophisticated level, it may well constitute the overarching influence dictating the course of graphic representation. Closely allied to these issues are the environmental conditions, ecological factors, and perceptual imperatives that nurtured the visual brain. By examining some of those criteria, the possibility arises of identifying the precursors that led to the first representations and how they relate to early human behaviour.

A guiding thread throughout this book concerns the detection of animals, especially as animals were crucial to the survival of palaeo-people during the Ice Age, particularly on hunting forays and during everyday activities when predators needed to be avoided. As a result, the ability to identify animals in challenging environmental situations involving camouflage can be regarded as one of the key factors that led to the human visual system's substantial capabilities.

This book takes a radically different approach to the human obsession with the arts over the long term, especially the depiction of animals. In pursuing this goal, I hope to persuade you of the relevance of visual neuroscience and neuroarchaeology to understanding the provenance of such inscrutable depictions. An approach to the origins of art from this perspective is viable thanks to the fascinating data deriving from the relatively new science of brain scans, which shows how the visual pathways are affected both by incoming perceptual information and the way this is assimilated across the brain

In order to explore how visual neuroscience and related disciplines can shed light on palaeoart, we apply a "bottom-up" approach beginning with an exploration of the ecological factors that shape the way the early visual cortex processes information in response to evolutionary constraints. Then we move up through the hierarchy of the visual brain towards more elaborate processes until finally reaching the topmost layer where consciousness meets the socio-cultural. Along the way, the relevance of the various findings will be applied to palaeoart ultimately to provide novel insights as to how, and possible why, art was first created. In contrast to symbolic approaches, on which most existing theories are based, this book is grounded on the principles of enactive or embodied cognition that sees human behaviour as closely tied to the material world without much, if any, cognitive mediation in the sense no separation between the subjective and objective is deemed to exist. However, this may be somewhat of a simplification, as how the world is experienced may initially stem from the way it is encoded and held in neural circuits. Thus, cognitive processing is inevitable in that the essentials of embodiment are subject to cognitive reassignment. It is how the world, as perceived, is reconfigured by the higher visual association areas that provides a fascinating new way of approaching the enigma of palaeoart.

Although identifying an object in a graphic image is often taken for granted, it belies a highly complex series of perceptual and cognitive events. In order to understand the means by which the first iconic images were produced, disentangling such complexity is essential. By so doing we will be in a much better position to address issues concerning how the first representational art arose. This is not an easy task as the brain did not evolve to understand itself rather it emerged from the need to deal with ongoing issues of survival and the complexities of social life. As graphic images exist as an extension of the brain, understanding how they first came about becomes even more problematic. In a way, pictures have a foot in both camps, the visual cortex (including the visuo-motor system) and the outer world, which makes graphic images all the more fascinating. It is often claimed, because of its remoteness in time, the reason why palaeo-humans began to create depictions of animals will forever remain a mystery. However, the issue of how and why individuals began producing images of animals is one that transcends archaeology and anthropology as it has implications for understanding the way the brain functions in relation to how "images" are constructed in the cortical systems. It is how the first figurative depictions fit into this broader context that provides a novel twist to this book.

## The visual brain on the African savannah

It is generally agreed that the reason the eye evolved five hundred million years ago during the Cambrian explosion was to improve the performance of organisms in detecting prey, thereby promoting successful predation (Parker, 2003; Ingram, 2002).[2] Although the eye eventually became an extremely complex organ, and is one of the "miracles" of the gradual course of evolution, in larger brained mammals the real task of vision takes place in the higher visual brain. Thus, despite the relative sophistication of the eye, we can be sure the visual brain is where real seeing takes place. As the visual areas of the primate brain are responsible for the actual phenomenon of seeing, it is safe to say that, as miracles go, this is an even more astounding evolutionary achievement than the eye itself. This might not seem obvious, as the act of seeing is taken for granted, so that, to the ordinary observer, seeing appears to take place in the eye itself. The fact so many areas of the brain are devoted to visual input suggests otherwise. In fact, the split second it takes to recognise an object belies a complex functional hierarchy whereby the visual image is formed and engages conscious awareness. As the visual faculty evolved over millions of years, and is vital to survival, many of its functions have become unconscious and automatic. The 150-300 ms it takes to recognise an object may give the impression of effortless perception, but the need to detect things in the world quickly and reliably has led to many of the underlying functions becoming automatic. Although this promotes fast reactions to potential threats, it does not describe the underlying neuro-visual processes. By examining research into this issue from the perspective of evolution—especially with regard to form perception—a story unfolds informing us of the very ecological niche in which our early ancestors lived and died.

We should not underestimate the task faced by the visual system during evolution. Think of what is involved in trying to decode the complicated signals that constitute the rich tapestry of the visual array, often referred to as the "blooming, buzzing confusion." In effect, the visual brain has to contend with a torrent of ambiguous information. For example, a particular arrangement of lines, as perceived, can be interpreted in an enormous number of ways. In fact, not even the fastest modern computers come anywhere near the capabilities of the visual system for tackling the problem (Cosmides and Tooby, 1994; Humphreys and Riddoch, 1992), though in the 21st century great strides have been made towards this. How, then, are we able to deal with the confusion emanating from the visual world when most of the perceptual mechanisms operate at a subliminal level?

One of the tricks the visual brain employs is modularity (Foder, 1983), as it not only operates separately to other sensory modalities, but contains subsystems and sub-subsystem that perform different tasks for dealing with raw incoming visual information. This is because it is easier and more efficient for a system to cope with a complex task by having separate specialists at each level rather than having a jack-of-all-trades but master of none. Hence, through modularity the brain is able to divide and conquer thereby simplifying the task in hand. Depending on prevailing circumstances, however, this modularity—especially in humans—may come under the influence of higher cognitive processes (Vetter and Newen, 2014).

In that way, evolution provided ancestral hunter-gatherers eking out an existence on the African savannah from around 300,000 years ago with an exquisitely tuned visual system that helped facilitate survival. How do we know the human visual cortex derived from evolutionary imperatives? One reason is that many of the same or similar visual regions of humans can be found in other primates, such as macaques and chimpanzees (Orban et al. 2004).[3] In addition, vision benefits from its own "intelligence" relatively independent of language, which allows things in the world to be identified and categorised according to inherent processes. From an evolutionary standpoint, it is commonly accepted that visual "intelligence" precedes linguistic understanding because it relies on tacit knowledge gained from long term, as well as ongoing, experience with the world (Hoffman, 1998).

The visual brain was "designed" by evolution in order to cope with the world as efficiently and effectively as possible. Each module is then able to bring to bear its own particular specialisation for the rapid assimilation of incoming data from earlier levels to be transmitted to later stages of the processing hierarchy. Indeed, it has been established that visual areas are not just modular in their different capacities at different layers of processing (horizontally) but also in the pathways that handle the flow of information from one level of this hierarchy to the next (vertically) (Livingstone and Hubel, 1987).

The visual cortex also has separate functional areas and streams that, amongst others things, encode such important attributes as form, movement, and colour. But, the story does not end there—a crucial challenge with which this system has had to deal concerns constancy. Constancy is defined as the ability to attend in predictable ways to information essential to humans as a species that is rendered stable in the face of an ever-changing visual array, thereby enabling signal to be discriminated from noise. As constancy is crucial to our later discussions on art, it is essential to bear this concept in mind.

The visual cortex consists of at least thirty different sub-regions (Orban et al. 2004). In addition, numerous areas partially process visual information throughout the brain, mainly by interfacing with corresponding sensory and associative systems. That complexity demonstrates the importance of vision to hominins even before the appearance of the anatomically modern human brain some 300,000 years ago. Thus, the brain has been shaped by recurring evolutionary events leading to expanded forward visual areas and frontal cortices.

The close-knit relationship between the more obvious attributes of the human visual system and the ancestral environment from which it derives is demonstrated in traits such as stereo vision. This is believed to have originated from the need to perceive insects at close quarters (Julesz, 1995). The decoding of camouflage may also have played a role, as form can be disentangled from distracting cues more efficiently through stereo vision (Allman, 1999). Significantly, front-facing eyes are thought to derive from our role as predators, which confers highly-focused seeing for accurate perception at a distance. One disadvantage of eyes arranged in this manner is that humans become prone to attack from the rear. As social animals, however, the eyes of others group members conveniently serve as a communal lookout in the event of danger.

## Surviving on the savannah

Hunter-gatherers from 300,000 thousand years ago originally descended from groups in Africa that had to contend with considerable climatic and environmental challenges during the Pleistocene (Scerri et al. 2018). Late Pleistocene European communities became vulnerable to additional threats, including the vagaries of an extremely cold climate, competing hominins i.e., Neanderthals, starvation, disease, and new predators. Survival depended on the particular faculties with which evolution had endowed humans. Amongst other things, these consist of an upright posture, bipedalism, the ability to sweat, lack of hair, a prodigious opposable thumb, and, of course, a large brain, to some extent fashioned according to the demands of vision (Vyshedskiy, 2014). Ancestral hunter-gatherers, however, remained unaware of the complex, highly tuned visual mechanism that allowed the world to be discerned. They would also have been unaware that this ability existed not to survey the beauty of the world, but rather because it promoted vigilance for the purpose of survival, as evolution functions only at the level of direct action with the world (Milner and Goodale, 1995). Extinction awaited any primate who regarded a plant or animal as a thing of beauty rather than a potential source of food or a threat.

The fact that an indistinct animal could pose a threat or, alternatively, serve as a source of food, meant that enhanced perceptual capacities were essential to promote accurate detection. Identifying an animal in degraded viewing conditions is, however, far from straightforward, as an animal can be seen from a number of directions with the profile changing drastically according to viewing angle. Obviously, learning that an animal is harmful from only one direction, say obliquely, when it might be approaching head-on, is ineffective—hominins would not have survived if this had been the case. Rather, animals are relatively easy to identify from most angles but this begs the question how this skill first came into being.

It transpires that constancy, the ability to recognise an object from a number of different views is, in itself, complex. Not only is it essential to recognise an object from unusual angles but also when size, translation, lighting/shadow, colour, all undergo change. Some of those problems are dealt with in the higher areas of the visual cortex while other more elementary aspects of form are decoded nearer the input end. The early perceptual stages deal with "basic" visual information such as contrast enhancement, edge and contour detection, grouping of common features, texture assimilation, orientation preference etc. In fact, all the way from the retina to the higher brain considerable enhancement of visual information takes place. At the retina, this involves organizing photons as points of contrast by coordinating visual information at the level of the smallest units. However, no image is actually perceived or subjectively experienced at the retina, only a scintillating pattern of spots of light. At the highest level—where a potential approaching big-cat is identified as such—the visual system sees the world in much larger manageable chunks. This is because, from the eye to the later stages, a successive augmentation in the ability to deal with extensive amounts of information occurs; a proclivity achieved through earlier levels in the hierarchy relaying the compiled results to subsequent stages (this process is reflected in the receptive fields, which are smaller in earlier compared to later stages).

Competing with large carnivores for prey was unavoidable during the Ice Age and, accordingly, led to an emotional response in the form of the delivery of massive amounts of adrenaline and corresponding neurotransmitters throughout the nervous system, readying the body for fight or flight. In fact, the emotional brain is alerted well before say a big cat is recognised by the higher visual cortex. How is that possible when the feline may not yet have been consciously identified? The answer is that conscious awareness is "emergent" in that it depends on implicit, preconscious mechanisms (Dehaene et al. 2006; van Vugt et al. 2018). There is, in fact, a pre-conscious route from the early visual centers to the emotional brain (in the limbic system) that is activated before explicit recognition occurs in the inferotemporal cortex and prefrontal areas (at the higher end of the visual cascade). This early warning system, which extends from the initial stages of the visual stream directly to the emotional areas, provides a fast track route that is sensitive to minimal cues for signaling danger (LeDoux, 1994, 2003). As the bodily systems need to be primed for immediate response should explicit recognition confirm what the emotional system suspects, this provides an important and effective safety switch. As it happens, the emotional system seems particularly sensitive to visual cues warning of danger, which has important implications for our discussion on palaeoart in later chapters.

One outcome of those deliberations is that the visual system, although wonderfully adapted for achieving constancy in an ever-changing world, acts as a filter through which incoming information flows. Remember that the visual world "out there" is an infinitely ambiguous commodity, to which the brain has had to adapt in order to extract particular cues critical to survival. This filtering begins at the retina where information, collected by the 101 million retinal cells, converges onto one million fibres of the optic nerve. In order to pass through the filter, information from the retina is extensively edited and condensed with no more than the crucial features allowed through. This means only those aspects of the visual array vital to humans are assimilated by the visual brain. Transmission of the entire visual flux of the world would not only be self-defeating but unnecessary. The visual brain can thereby be regarded as a kind of information cruncher, molded by information that has, over the generations, promoted successful reproduction. It comes as no surprise then, that, in order to pack so much computational expertise into a diminutive brain, the cortex looks like a large walnut, the many folds serving to boost storage-capacity. Yet, despite the millions of neurons and the accompanying billions of interconnections, the visual brain can still be regarded as a kind of filter. Fundamentally, the kind of visual information to which we have become attuned as a species during evolution has determined the information that passes through this filter.

## Seeing the same difference

Competition with large carnivores for the same habitat has been implicated as a determining factor in human evolution, especially as we needed to outwit predators through increased intelligence (Brain, 1981; Camarós et al. 2016), which, again, was probably one of the main factors contributing to brain expansion and complexity. Human interaction with animals constitutes an important segment of the hominin evolutionary niche that is reflected in the diverse ways in which various species interact with their habitat by responding to different kinds of visual stimuli. For example, compared to humans, birds and insects are sensitive to a different part of the spectrum (short wavelength), so they see "colors" and patterns in flowers to which we remain oblivious. This is because their nervous systems evolved to be sensitive to such cues as part of a symbiotic relationship with plants. Not only have plants provided nectar for insects as a reward for distributing their pollen but also made sure, by furnishing eye-catching signals, they are easy to spot and locate. Accordingly, insects stood a better chance of passing on their genes. Correspondingly, the flower that attracts the most insects through the best signaling device stands a better chance of spreading its pollen and increasing reproduction. A similar dynamic probably determined the sensory abilities of humans according to the particular environmental factors crucial to survival.

In the case of hominids, evolutionary theory suggests full colour competency originally derived from when primates became daytime rather than night-time foragers. That is, when ripe fruit and berries became a prominent part of the diet rather than mere leaves and insects (Mollon, 1995). Red-green colour blindness illustrates the point, as individuals with this form of blindness have difficultly detecting various kinds of ripe fruit in leaf cover. Indeed, similar to how insects co-evolved a sensitivity to particular wavelengths of light, it is believed plants, especially fruit-bearing trees, shaped human colour sensitivity.

The reward for our early ancestors took the form of sugar and other nutrients in fruit and berries, of which the seeds and pips passed through the digestive tract to be later deposited at a distance as natural manure for later germination. Larger fruit, taking longer to eat, might have been carried to distant locations, where the inedible stones or kernels would have been discarded on the ground. So, we see that the capacity for perceiving colour was honed by the ecological context that specified the spectrum to which we became sensitive. Thus, the environment led to important evolutionary adaptations that stemmed from the relationship between fruit bearing plants and the need for sustenance. In sum, a mesh occurred between the prevailing environmental conditions, on the one hand, and the human visual system on the other, leading to the emergence of particular sensitivities. Another important evolutionary constraint illustrating how adaptive traits formed concerns the human immune system, which evolved from a long-term interaction with bacteria and viruses that culminated in various types of antibodies as part of the human immunological defense system.

Given the way colour vision evolved, is it possible to similarly "backward engineer" the evolutionary precursors for how we came to perceive form? This is perhaps a more fundamental attribute than the ability to see colour, as it would have been almost impossible for a colour blind person to survive on the Serengeti without the ability to see form.[4]

What key evolutionary factors might thereby be responsible for the perception of form? Suppose a big cat is bounding towards one of our ancient ancestors, that individual will have long realised this animal means business and has already taken aversive action by, for example, climbing the nearest tree. But the carnivore would have been stalking for some time, waiting for the right moment to pounce. In which case, an upright posture allows a human look-out—a posture most other primates lack—to peripherally detect movement in the grass. Such an upright stance also comes with the advantage of eyes perched on a tall bony scaffold allowing distant objects to be viewed. The lurking carnivore, however, is aware of this and keeps low to the ground advancing in short, careful, movements. Despite the benefit of camouflage, the predator instinctively knows movement increases the chance of detection in that its upper outline is more apparent against the surrounding vegetation. Movement is such a vital clue to impending danger that peripheral vision is highly sensitive to its occurrence to the extent it has become pre-attentive in orienting the other sensory modalities. The importance of movement is emphasised by the fact a distinct, modular processing stream exists from the retina to the higher visual cortex (known as the magnocellular pathway). Movement thereby provides an initial indication as to the presence of a predator, tagged first by peripheral vision that brings focal vision to bear on the object in question. This is borne out by rods in the retina that specialise in detecting movement—also employed for night vision—that have a longer evolutionary history than the cone system (fovea/colour), which underlines the importance of the perception of movement for survival.[5]

The stalking carnivore, however, might notice such an orienting response and again becomes a rigid non-entity, and the human look-out surmises the disturbance was caused be a sudden breeze or perhaps a small rodent in the undergrowth. It is only when figure and ground become differentiated that the big-cat is unequivocally identified.

The human perceptual "kit-bag" contains further tricks to aid detection. One such trick concerns perceptual priming whereby certain features useful for detecting animals allow anything resembling an animal to be rapidly discerned. The visual system, for example, is particularly adept at detecting contours/edges as well as curved outlines (Gibson, 1933). Those important perceptual "primitives" constitute the bounding contour of an animal, vital for identifying a partially hidden animal. In addition, the preference of the mid-level visual system for good continuation of any disrupted contour means that an outline, which might be attenuated by camouflage, can be discerned. Such capacities derive from the need to achieve constancy in a variety of everyday situations. Interestingly, humans are more able to discern global structure from local features compared to other higher primates (Fagot and Tomonaga, 2001), which derives from an enhanced visual system honed by the need to detect animals in various situations (Vyshedskiy, 2014).

Another scenario where constancy is useful is in subdued lighting. A special mechanism comes into play here, where the recognition system engages in hypotheses-testing by providing options when confronted with uncertain stimuli. Indeed, visual memory may have derived from such a capacity in the sense that the ability to summon images in the "mind's eye" seems to depend on rules employed by the recognition system (Shepard, 1984; Kosslyn et al. 1990; Farah, 2000), thus providing a means to manipulate imagery (Kirby and Kosslyn, 1992). Imagery, however, is prone to fading especially as it lacks updates from immediate sense data, which can give rise to problems with information retrieval and recall. Whether what is experienced visually derives directly from the world or visual memory, this has implications for understanding how artists represent the world.

Many of the perceptual ruses outlined above derive from the need to detect predators or potential prey in "noisy" visual environments. Through camouflage and related strategies, a stalking predator seeks to ensure the environment remains unchanged. In other words, every time a hunter observes what might appear to be an animal, the tracker is duped by the predator into thinking grass, grass, grass, or foliage, foliage, foliage, when, in actual fact, there may well be grass, leopard, grass, or foliage, lion, foliage. This game of hide and seek persisted throughout evolutionary time whereby the human perception/recognition system was shaped by predators targeting hominins. In this way, as much as a predator's beguiling strategies became elaborate, so hominin visual capacities became ever more sophisticated.[6]

The perceptual skills for detecting a partially hidden, degraded, or camouflaged carnivore can be exploited for tracking ruminants and ungulates. Those skills were fundamental to survival because the stalking and capture of animals for consumption provided the necessary protein for brain expansion (Martin, 1983); an organ that requires vast amounts of sustaining energy. As Allman (1999) indicates, larger brains evolved from the need of predators to remain competitive by processing greater amounts of information with increased proficiency. In becoming a top predator, this necessitated that humans engaged in group cooperation, which required a protracted training regime during development (Levy, 1999).

When the visual system becomes emotionally aroused, vigilance towards danger tends to increase, thereby enhancing alertness. That mechanism, however, can become dysfunctional when a traumatic event is experienced, where the original event tends to spontaneously spring to mind (in extremis, referred to as post-traumatic stress disorder). This occurs because unanticipated events lead to a rush of adrenalin that is slow to dissipate, thus helping perception remain primed to any potential danger, which makes sense, as when a carnivore attacks it is likely to loiter with intent—a hunter will be attempting to temporarily remain hidden in the immediate vicinity and big-cats know this. We also need to bear in mind that carnivores and prey were considerably larger and more powerful than modern equivalents leading to sustained levels of arousal. The associated adrenaline surge thereby allows perceptual thresholds to be lowered by facilitating increased sensitivity to any visual cue signaling an animal. An observation that is in line with a conservative evolutionary approach, where it is a safer to respond to an unlikely cue than not because the costs of not responding to a potential cue is much greater than over responding—a better-safe-than-sorry-strategy (Van de Cruys and Wagemans, 2011).

Animals were also important as a source of fuel and clothing—even their bones were exploited for making specialised tools and free-standing shelters. Animals therefore played a crucial role in hunter-gatherer lifeways but it was the predators competing for the same quarry that were especially important.[7] In addition, the tools of hunter-gatherers were mainly used to defend against predators and for the killing and butchery of prey (Bunn, 1981; Potts and Shipman, 1981; Shipman and Rose, 1983, Shipman, 2010). The making and use of tools, however, dates to two to three million years ago, before hominins became proactive hunters, when they were restricted to scavenging carcasses abandoned by large carnivores. The first crude stone tools were therefore utilised to provide access to the nourishing marrow contained in the remains of robust animal bones.

Mayr (2002) suggests that the increase in brain size from Australopithecus to later *Homo* occurred because of migration from a dense tree environment to an open bush savannah. As the savannah was unable to provide enough cover from carnivores, hominins needed to devise strategies to avoid predators. One of the more important strategies centred on better communication through speech, which enhanced the planning and organisation of groups during hunting expeditions (Lombard, 2015). If possible, it is beneficial to consign long-term and reliably persistent properties of the environment to the hard-wired architecture of the brain. This allows information to be stored at the level of genes through genetic assimilation[8] so that the relevant neuronal substrates for encoding important visual stimuli, such as faces and animals, are engaged at an early age. For example, three-year-olds have an uncanny ability to spontaneously classify animals according to basic categories without previous instruction (Gellman, 1990), a topic to which we will return below. Consigning such criteria to genetic markers means each new generation does not have to learn anew, with all that this entails, the uncertainties and dangers suffered by previous generations. Infants therefore have the advantage of being pre-adapted to potential dangers, although, as the child matures, that ability is refined through learning. Related to this, sensitivity to movement—an indicator of potential danger—is thought to be present at birth.

But what has all this got to do with the origins of representational art? In the following chapters, the relevance of the forgoing will become obvious. First, and in order to set the scene as to how art figures in this story, we need to delve further into the inner workings of the visual brain that will reap rich rewards for our later discussions on palaeoart.

## Notes

1. The medial orbito-frontal cortex has also been associated with encoding aesthetic criteria (Ishizu and Zeki, 2011)—including an adjacent area—the right orbito-frontal cortex (for the emotional judgments of artworks but also active when observing faces and scenes). Also active when subjects judged abstract artworks as emotionally engaging—especially when linked to "goodness"—is the inferior frontal gyrus (Melcher and Bacci, 2013). Those areas may be interconnected in the sense that the orbito-frontal cortex is associated with person perception and the inferior frontal gyrus with positive affect through connections with the limbic system—where the good and beautiful are associated with empathy, reward and social judgements—suggesting a link to social affiliation. The anterior insula, also associated with aesthetic experience, may be similarly involved in this network (Brown et al. 2011) in that it is also related to empathising (Singer, 2006). As an integrated system for assessing the saliency, value and significance of objects—both social and asocial—for promoting survival, such a network seems to have formed the conduit out of which cultural artefacts could eventually flow (Brown et al. 2011).

2. Parker puts forward "The Light Switch Theory" based on the idea that the first eyes of the Cambrian explosion evolved as a consequence of adaptive pressures to do with the interaction of predators and prey. Ingram holds the view that the eyes of creatures as diverse as flatworms, arthropods, cephalopods, and vertebrates appear to be derived from a prototypical light-sensing organ that was present in the common ancestor of all groups that existed some 550 million years ago.

3. This debate, however, continues to rumble on with those who still hold that human intelligence and consciousness constitutes a step change that makes humans largely independent of the determinants to which the rest of the animal kingdom is

prone. This implies that what makes humans different is still an indefinable quantity but, nevertheless, is probably related to raised awareness associated with enhanced consciousness due to the fact that the higher-order faculties that make humans human are not totally dependent on the lower-order, more modular, structures typical of mammalian brains.

4. Paradoxically, during the Second World War, individuals suffering colour blindness were employed by the Royal Air Force because they were able to discern certain kinds of camouflaged objects more easily than those with normal visual abilities.

5. The cone system is thought to have originally evolved from a mutation deriving from the colour blind rod cells and is therefore much more ancient than cones. Rods were, therefore, tuned to movement before cones came along.

6. In a cave complex at Swartkrans, South Africa, the remains of 130 Australopithecine individuals were found along with those of carnivores and herbivores. It was discovered that the holes in the cranium of a juvenile Australopithecine matched the canines of a leopard from the same cave, suggesting that hominins were hunted just as they themselves were on the road to becoming hunters (Brain, 1981). Brain found that leopards dragged their victims into trees where they were devoured, safe from other predators and scavengers such as hyenas.

7. Particularly important would have been big-cats, hyenas, bears and wolves not forgetting that the large herbivores such as mammoths, bison, and aurochs also posed a real threat.

8. This idea was put forward by Conrad Waddington (1957) who stated that natural selection tends to replace flexible adaptive responses to continuing environmental constraints with genetic predispositions. The consequence of this is that flexible responses during development gradually become more constrained by genetic inheritance, responses that give rise to predispositions towards particular environmental stimuli. Making such responses mandatory provides for a more efficient and less risky behaviour than one that relies on each individual having to learn the relevance of every environmental signal from scratch.

## References

Allman, J. M. 1999. *Evolving Brains.* New York: Scientific American Library.

Boyd, B. 2005. Evolutionary theories of art. In, J. Gottschall and D. S. Wilson (Eds.), *The literary animal* (pp.149-178). Evanston: Northwestern University Press.

Boyd, R. and Richerson. P. J. 2005. *The Origin and Evolution of Cultures.* New York: Oxford University Press.

Boyd, R. and Richerson, P. J. 2007. Cultural adaptation and maladaptation: Of kayaks and commissars. In, S. Gangestad and S. J. Simpson (Eds.). *The evolution of mind: Fundamental questions and controversies* (pp.327-331). New York: Guilford.

Brain, C. K.1981. *The Hunter or the Hunted?* Chicago: University of Chicago Press.

Brown, S. and Dissanayake, E. 2009. The arts are more than aesthetics: Neuroaesthetics as narrow aesthetics. In, M. Skov and O. Vartanian (Eds.). *Neuroaesthetics.* (pp. 43-57). Amityville: Baywood.

Brown, S., Gao, X., Tisdelle, L., Eickhoff, S. and Liotti, M. 2011. Naturalizing aesthetics: Brain areas for aesthetic appraisal across sensory modalities. *Neuroimage.* 58: 250-258.

Bunn, N. 1981. Archaeological evidence for meat-eating by Plio-Pleistocene hominids from Koobi Fora and Olduvai Gorge. *Nature.* 291: 547-577.

Camarós, E., Cueto, M., Lorenzo, C., Villaverde, V. and Riva, F. 2016. Large carnivore attacks on hominins during the Pleistocene: a forensic approach with a Neanderthal example. *Archaeological and Anthropological Sciences.* 8(3):635–646.

Conard, N. J. 2008. A Critical View of the Evidence for a Southern African Origin of Behavioural Modernity. *South African Archaeological Society Goodwin Series.* 10: 175–9.

Cosmides, L. and Tooby, J. 1994. Origins of domain specificity: The evolution of functional organization. In, A. Hirschfeld and S. A. Gelman (Eds.), *Mapping the Mind. Domain specificity in cognition and culture* (pp. 85-116). New York: Cambridge University Press.

Davies, S. 2012. *The Artful Species: Aesthetics, Art, and Evolution.* Oxford: Oxford University Press.

Dehaene, S., Changeux, J-P.,Naccache, L., Sackur, J. and Sergent, C. 2006. Conscious, preconscious, and subliminal processing: A testable taxonomy.*Trends in Cognitive Sciences.* 10 (5):204-211.

Dissanayake, E. 2000. *Art and intimacy: How the arts began.* Seattle and London: University of Washington Press.

—. 1999. "Making Special": An undescribed human universal and the core of a behavior of art. In, B. Cooke and F. Turner (Eds.). *Biopoetics: Evolutionary explorations in the arts* (pp.27-45). Lexington, KY: ICUS.

Dutton, D. 2009. *The Art Instinct.* Oxford: Oxford University Press.

Fagot J. and Tomonaga, M. 2001. Effects of element separation on perceptual grouping by humans (*Homo sapiens*) and chimpanzees (*Pan*

*troglodytes*): Perception of Kanizsa illusory figures. *Animal Cognition.* 4:171–177.
Farah, M. J. 2000. *The Cognitive Neuroscience of Vision.* Malden, Mass.: Blackwell.
Fodor, J. 1983. *The Modularity of Mind.* Cambridge, Mass.:MIT Press/ Bradford.
Gelman, R. 1990. First principles organize attention to and learning about relevant data: Number and the animate-inanimate distinction as examples. *Cognitive Science.* 14: 79-106.
Gibson, J. J. 1933. Adaptation, after-effect and contrast in the perception of curved lines. *Journal of Experimental Psychology.* 16:1-31.
Hodgson, D. and Verpooten, J. The evolutionary significance of the arts: Exploring the byproduct hypothesis in the context of ritual, precursors, and cultural evolution. *Biological Theory.* 10(1): 73-85.
Hoffman, D.D. 1998. *Visual Intelligence.* New York: W.W. Norton.
Humphreys, G. W., Riddoch, M. J. and Boucart, M. 1992. The Breakdown Approach to Visual Perception: Neuropsychological Studies of Object Recognition. In, G. W. Humphreys (Ed.), *Understanding Vision, An Interdisciplinary Perspective* (pp.104-125). Oxford: Blackwell.
Ingram, J. N. 2002. A possible Precambrian origin for topographic maps in the visual system. *Trends in Neuroscience.* 25 (5): 232.
Ishizu , T. and Zeki, S. 2011. Toward A Brain-Based Theory of Beauty. *PLoS ONE.* 6(7): e21852.
Julesz, B. 1995. *Dialogues on Perception.* Cambridge, MA: MIT Press/ Bradford.
Kirby, K. N. and Kosslyn, S. M. 1992. Thinking Visually. In, G. W. Humphreys (Ed.), *Understanding Vision* (pp. 71-85). Cambridge, MA: Blackwell.
Kosslyn, S. A., Flynn, R.A., Amsterdam, J. B. and Wang, G. 1990. Components of High Level Vision: A cognitive neuroscience analysis and accounts of neurological syndromes. *Cognition.* 34: 203-277.
LeDoux, J. E. 1994. *Emotion, Memory and the Brain. Scientific American* (June): 32-39.
—. 2003. The Emotional Brain, Fear, and the Amygdala. *Cellular and Molecular Neurobiology.* 23:4/5:727-738.
Levy, C. K. 1999. *Evolutionary Wars.* New York: W. H. Freeman.
Livingstone, M. S. and Hubel, D. H. 1987. Psychophysical evidence for separate channels for the perception of form, colour, movement, and depth. *Journal of Neuroscience.* 7: 3416-3468.

Lombard, D. 2015. Hunting and Hunting Technologies as Proxy for Teaching and Learning during the Stone Age of Southern Africa. *Cambridge Archaeological Journal.* 25(4): 877-887.

Martin, R. D. 1983. Human Brain Evolution in an Ecological Context. *The Fifty-second James Arthur Lecture on the Human Brain.* New York: American Museum of Natural History.

Mayr. E. 2002. *What Evolution Is.* London: Phoenix.

Melcher, D. and Bacci. F. 2013. *Perception of Emotion in Abstract Artworks: A Multidisciplinary Approach. Progress in Brain Research.* 204: 191-216.

Miller, G. F. 2001. Aesthetic fitness: How sexual selection shaped artistic virtuosity as a fitness indicator and aesthetic preferences as mate choice criteria. *Bulletin of Psychology and the Arts.* 2(1): 20-25.

Milner, A.D. and Goodale, M.A. 1995. *The visual brain in action.* Oxford: Oxford University Press.

Mollon, J. 1995. Seeing Colour. In, T. Lamb and J. Bourriau (Eds.), *Colour: Art and Science. The Darwin College Lectures.* (pp. 127-150). Cambridge: Cambridge University Press.

Orban, G. A.,Van Essen, D. and Vanduffel, W. 2004. Comparative mapping of higher visual areas in monkeys and humans. *Trends in Cognitive Sciences.* 8(7):315-324.

Parker, A. 2003. *In the Blink of an Eye.* Cambridge, Mass.:Perseus.

Potts, R. and Shipman, P. 1981. Cutmarks made by stone tools on bones from Olduvai Gorge, Tanzania. *Nature.* 291:577-80.

Richerson, P. J. and Boyd, R. 2001. Built for speed, not for comfort: Darwinian theory and human culture. *History and Philosophy of the Life Sciences.* 23: 425-465.

Scerri, E. M. L., Thomas, M. G.,Manica, A., Gunz, P., Stock, J. T., Stringer, C., Grove, M., Groucutt, H. W. et al. 2018. Did Our Species Evolve in Subdivided Populations across Africa, and Why Does It Matter? *Trends in Ecology & Evolution.* doi.org/10.1016/j.tree.2018.05.005

Shepard, R. N. 1984. Ecological Constraints on Internal Representation: Resonant Kinematics of Perceiving, Imagining, Thinking and Dreaming. *Psychological Review* 91 (4):417-447.

Shipman, P. 2010. The Animal Connection and Human Evolution. *Current Anthropology.* 51(4):519-538.

Shipman, P. and Rose, J. J. 1983. Early hominid hunting, butchering and carcass-processing behaviours: approaches to the fossil record. *Journal of Anthropological Archaeology.* 2: 57-98.

Singer, T. 2006. The neuronal basis and ontogeny of empathy and mind reading: review of literature and implications for future research. *Neuroscience and Biobehavioral Reviews*. 30 (6): 855–63.

Van de Cruys, S. and Wagemans, J. 2011. Putting reward in art: A tentative prediction error account of visual art. *i-Perception.* 2: 1035–1062.

van Vugt, B., Dagnino, B., Vartak, D., Safaai, H., Panzeri, S., Dehaene, S., Roelfsema, P. R. 2018. The threshold for conscious report. *Science.* 360: 537–542.

Vetter, P. and Newen, A. 2014. Varieties of cognitive penetration in visual perception. *Consciousness and Cognition.* 27: 62–75.

Vyshedskiy, A. 2014 (2nd edit). On the Origin of the Human Mind. *MobileReference.* http://mobilereference.com/mind/online/index.htm.

Waddington, C. H. 1957. *The Strategy of the Genes*. London: Allen and Unwin.

Zaidel, D. W. 2010. Art and brain: insights from neuropsychology, biology and evolution. *Journal of Anatomy.* 216:177–183.

# CHAPTER TWO

# THE ECOLOGICAL CONTEXT OF THE VISUAL BRAIN

> Non-human animals were predators on humans; food when they strayed close enough to be worth pursuing; dangers when surprised or threatened by virtue of their venom, horns, claws, mass, strength, or propensity to charge; or sources of information about other animals or plants that were hidden or occluded. (New et al. 2007, p. 16598)

## Pre-conscious cues and the emotional brain

Since hominins separated from other primates around six million years ago and subsequently began to descend from the trees to eventually live on the open savannah, they inevitably became prone to large predators with which they competed for the same quarry. In order to survive in the new conditions, exceptional visual capacities were essential to detect predators at a distance, which also became indispensable for detecting prey; prey that became increasingly important to facilitate brain expansion (Aiello and Wheeler, 1995). Acute awareness of animals in various situations was therefore essential (DeCasien et al. 2017); an awareness that went hand in hand with refined hunting skills (Hart and Sussman, 2009; Brain 1981). In order to appreciate how this unfolded, we need to identify the determining visual environment that shaped and gave rise to certain biases of the human visual system, as visual perception is predicated on how one engages with the world (Gilbert, 1996; Hodgson, 2004; Shepard, 1984). Such biases will have originally derived from immediate experience as well as certain longstanding evolutionary factors; one of which we saw concerns the need to find and detect animals in order to survive. In this chapter, we will be tracing the pathways in the visual brain that facilitated the ability of our early hominin ancestors to identify and discriminate animals. We begin with the subcortical emotional pathway.

There are two routes in the human brain for detecting a dangerous object which, during the Pleistocene, often involved animals—one cortical and explicit (the "high road"), the other subcortical (and subconscious, the "low road"), with the subconscious route being the quickest and most direct. LeDoux (1994), for example, has established that the subcortical system, involving the thalamus (a kind of relay station) to amygdala pathway, of both humans and animals responds similarly to potentially threatening stimuli, especially predators (an associated subcortical structure, the pulvinar, is also implicated [Pessoa and Adolphs, 2010]). This pathway is especially sensitive to animals as well as the human face (Yang et al. 2005; Martin, 2007). The subcortical pathway, however, furnishes only a crude rendering of the external world, whereas the cortical route delivers a more detailed representation. The subcortical route triggers an emotional response to a stimulus before conscious recognition ensues by priming fight or flight behaviour because not reacting to a stimulus incurs more costs than responding and, consequently, this pathway is particularly fast whereas the cortical route is relatively slow. As LeDoux (1994) makes clear, certain key features may be enough to trigger an emotional response via the limbic system, which includes the amygdala, before an object—such as a dangerous animal—is recognised. Thus, there seems to be an early warning system that proceeds directly from the retina, via the thalamus and pulvinar, to the limbic system for the rapid discrimination of potentially threatening objects that primes conscious awareness for action (Figure 2-1).

The brain may therefore simply store basic cues that can be verified by a delayed higher recognition system. For example, in a situation where we know snakes might be lurking, the sight of a slender curved shape on a path is sufficient to evoke a fear response, even though, on second glance, this may turn out to be a tree root (Figure 2-2). Initially, it is unnecessary to undertake a detailed analysis of the potential snake's attributes as these are irrelevant and would be detrimental to survival. So, the brain only needs to store simple cues that promote an immediate defensive response. Only when such crude information reaches the higher cortex is a stimulus consciously identified as such. In effect, the two systems operate in unison to provide a seamless conscious experience. A similar alerting mechanism has been identified for spiders, as stated by New and German (2015, p.10).

> The defining visual configuration of spiders appears to be the radiation of multiple segments from a central mass point with "legginess" being the most frequently reported frightening feature by spider-fearful individuals.

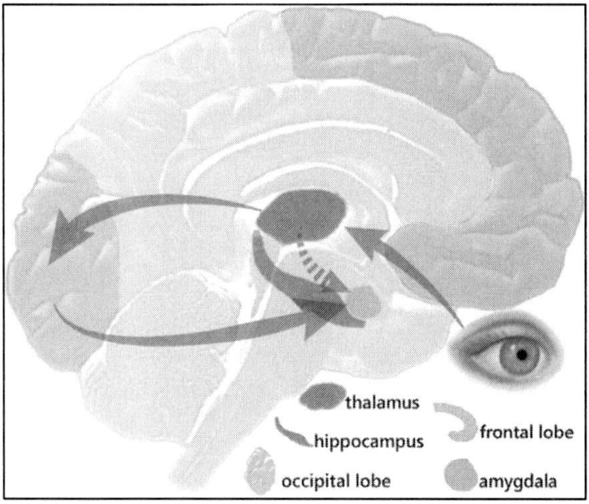


Figure 2-1. The fast low-road (dotted arrow) and "slow" high-road (complete arrows) neural channels of the visual brain. (Image by Manos Kefalas, Creative Commons Attribution-Share Alike 3.0 Unported)

In support of that observation, rigorous experiments demonstrate that we often respond to scary pictures of snakes and spiders even when not consciously aware of them (Koch and Crick, 2001)—a response that is extremely rapid and which exists in infants as young as 14 months as well as in nonhuman primates (Masataka et al. 2010; Isbell, 2006; Vida and Behrmann, 2017; Gomes et al. 2017).

In short, the low-road alerting system produces a generalised attentional bias for identifying threatening stimuli in a way that persists when activated (Ohman and Soares, 1998). This procedure encourages the visual system to concentrate on the cause of concern almost to the total exclusion of any other stimulus, which leads to a further entrenchment of anxiety/emotional arousal and a concomitant increased motivation to remain vigilant to any similar threat (a mechanism known as adaptive conservatism). Crucially, LeDoux (1998) found that, in most animals, more neural connections project to the higher cortical centres from the amygdala than the reverse, though humans benefit from greater top-down control compared to other primates. A finding that suggests the amygdala is primed to detect stimuli such as fauna (has lower trigger thresholds for such forms) that is translated into a bias towards such features at higher levels (see Figure 2-2).

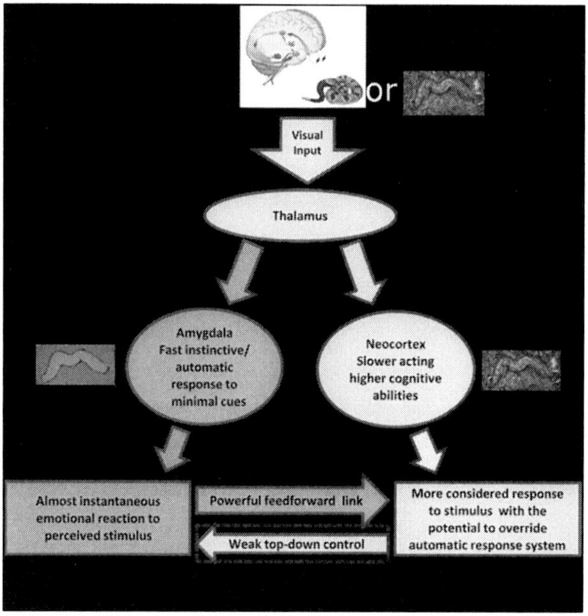


Figure 2-2. The fast and slow response pathways in the visual brain to a perceived fearful stimulus. In humans, there is greater top-down feedback and control to the amygdala compared to nonhuman primates (grey dotted rectangle). (Brain inset illustration in public domain).

In the same way as the above-mentioned "slender curved shape" or “legginess” evokes a fear response anticipating a snake or spider respectively, so a particular set of curved lines may facilitate the alerting signals in the emotional brain that code for predators, a hypothesis to which we will return below.

It seems, then, that the visual brain contains certain in-built capacities that allow for early and rapid emotionally-biased responses to particularly important stimuli crucial to survival. In this way, archaic humans became biologically predisposed to certain kinds of learning that led to particular brain modules specialised for perceiving natural forms such as animals (Caramazza and Mahon, 2003). The stimulus properties of those perceptual responses thereby constituted the foundation on which more sophisticated recognition processes were built.

The alerting mechanism for predators most likely generalised to those species upon which humans preyed because palaeo-humans were in competition with carnivores for the same quarry as well as needing to discriminate predator from prey. The alerting mechanism for attending to animate objects is referred to as "the animate monitoring hypothesis" (Tooby and Cosmides, 2005), which has received considerable empirical support. The following statement encapsulates the findings:

> We suggest that higher sensitivity for animate changes occurs because the animal has greater perceptual salience relative to other visual information. Our perceptual/cognitive systems appear to need less information from the environment to allocate attentional resources towards animate things. (Altman et al. 2016, p. 312)

In sum, human attention evolved in such a way so as to dependably and rapidly identify certain category specific features, especially animals and humans. Such categories elicit immediate but robust attentional resources compared to things such as mountains, plants, or manufactured objects.

Although preconscious cues are invariably abbreviated, occasionally processing at this level can be quite elaborate. For example, research suggests individuals can detect and categorise complex visual features of animals even when attention is focused elsewhere (Li et al. 2002). As a first line of defense we therefore rely on relatively simple key cues encoded in the emotional brain, which are preconscious and evolutionarily defined that facilitate a rapid response to potential threats. As these cues provide the first line of defense for detecting predators, it is not surprising they came to feature so widely in palaeoart.

Although preconditioned subcortical cues serve as an early warning system for signaling danger, other factors—to do with the immediate environmental context—helped to accentuate this process. As palaeo-humans were in daily contact with animals, the implicit (pre-conscious) and predetermined (evolutionary-defined) fundamentals underwent further improvement, known as "contingent capture" where top-down information increases the speed and efficiency of pre-attentive processing (Folk and Remington, 2005). So much so, that important features are cued implicitly as the following makes clear:

> The human visual system can quickly and efficiently extract categorical information from a complex natural scene. The rapid detection of animals in a scene is one compelling example of this phenomenon, and it suggests the automatic processing of at least some types of categories with little or no attentional requirement (Zhu et al. 2016, p. 1)

Due to the widespread feedback projections from the visual pathway to the emotional brain, there is an incremental elaboration of essential features of a perceived object until it is identified consciously (Cahill et al. 1996). In this way, a successive augmentation of the "image" occurs as information flows through the visual hierarchy from the most salient, enduring aspects to the fully embellished form.[1]

Objects frequently experienced, such as animals, are easily consigned to long-term memory through short-term visual recall, because this is reciprocally connected to the emotional brain (amygdala) via the hippocampus. Furthermore, confrontations with animals are highly charged, and at times traumatic, leading to a raised state of arousal and increased awareness, whereby the sympathetic system (part of the autonomic network) discharges adrenaline into the nervous system to promote evasive action. On one hand, palaeo-humans needed to approach quarry to expedite the hunt whereas, on the other, increased vigilance was required to detect predators stalking the same prey—a classic approach-avoidance scenario. With a visuo-emotional system readily prepared to detect animals, palaeo-humans were well equipped to survive in a dangerous environment. In fact, at basic and more complex levels, the visual system appears to act as a predictive coding device modulated by the emotional brain depending on context (Van de Cruys and Wagemans, 2011). Moreover, when the emotional brain is activated, memory tends to become more robust and alert to the cause of the activation.

The lessons from the inner workings of the visual brain in this section will provide a robust platform on which to disentangle the enigma of Palaeolithic art in the coming chapters.

## Minimal cues

We saw how perceptual alertness can be enhanced by increased activity in the emotional brain leading to heightened preparedness. It would, if possible, be advantageous to increase the efficiency and rapidity of this response. How can this be achieved? One way might be to pare down the visual signals to the absolute essentials so that only certain key features automatically alert the perceptual/recognition systems. We have already mentioned lines of curvature that serve as an early warning indicator to the presence of an animal. Having a set of key features available in visual memory closely linked to the emotional brain provides a head-start should a hasty retreat be required. A means of remaining alert to informative cues to guarantee safety is therefore essential. The question arises as to how those key features were assigned and how they figure in the larger scheme of things.

There are two ways this question can be addressed. The first relates to the notion of preparedness deriving from our ancestors' combined evolutionary histories whereby essential environmental cues became hard-wired in the visual brain. The second concerns how preparedness is associated with fixed action patterns (FAPs) and sign-stimuli discovered by Tinbergen and Lorenz (Lorenz and Tinbergen, 1938; Tinbergen and Perdeck, 1950; Tinbergen, 1951). As a result of adaptive evolutionary factors, FAPs tend to be in-built and give rise to stereotyped behaviour (Lea, 1984) with each species having a distinctive repertoire of FAPs (Hinde, 1982). Sign-stimuli (sometimes referred to as key-stimuli) act as releasers, and are environmental cues highly significant to an organism's survival that trigger FAPs. Evolution seems to have "decided" that to counter camouflage a raised sensitivity to key-stimuli provides an efficient means of promoting survival.

Alcock (1998) defines that process as stimulus filtering whereby a nervous system responds selectively by extracting relevant information from the sensory barrage, while, at the same time, ignoring most other things. Here, organisms "zero in" on biologically significant entities such as parents, sexual signals, predators or prey. Thus, different species benefit from different neural mechanisms for performing such tasks as dictated by the prevailing ecological niche. Such cues can be observed in animals, from "lower" to "higher" species. For example, herring gull chicks beckoning for food respond to the orange spots located on the mother's beak. During mating, newts respond to certain bodily signs. Frogs react to a natural stimulus that is either long and thin moving in one direction or a spot moving across the field of vision (usually a worm or insect, yet these cues elicit a response even when the object concerned is inappropriate).

Similar behaviour has been observed in primates where chimps react with a particular distress call when seeing a cross-shaped eagle-like form but not when this is goose-like (with a longer head/neck configuration). Similarly, east African vervet monkeys are preyed upon by leopards, two species of eagle, and pythons to which they react with different alarm calls. The call for leopards sees the simians charge off up into the trees, whereas the alarm for eagles results in looking upwards and a rush to the nearest bushes. The call for pythons, however, induces stasis with the vervets standing upright accompanied by much peering into the grass (Seyfarth et al. 1980). Such alarms are triggered by minimal visual (or auditory) cues that serve as a reliable early warning for the presence of different predators.

The important question is whether analogous stimuli might be relevant to humans? One stimulus relates to the parental attachment response, where facial features in infants are larger (e.g., the eyes), rounder or flatter than in adults (Gardner and Wallach, 1965) that induce a protective reaction in parents. Moreover, because cues to emotional status in the human face are fundamental to communication, research has found that individuals can discern whether a face is happy or angry even when presented subotimally (Murphy and Zajonc,1993). In other words, an individual can register key cues about facial expression that influence subsequent behavior in a way unbeknown to the person concerned. In addition, some investigators (Dziurawiec and Deregowski, 2002) have established that infants and adults are highly sensitive to eye-spots due to the importance of eyes in human communication, which are also crucial in detecting predators, especially at night. Another cue concerns the observation that infants become alarmed when confronted with looming objects, which are suggestive of a rapidly-approaching dangerous entity. Crucially, we have already established that the human visual system processes basic perceptual information of snake-like contours faster than other stimuli. This is accomplished by a fast feedforward sweep in the visual system at level of the primary visual cortex and beyond by way of base as opposed to incremental grouping (Schmidt and Vankleef, 2016). Interestingly, humans respond to curved contours more rapidly than nonhuman great apes (Munar et al. 2015), which may relate to the fact that curved outlines were crucial to early humans for perceiving animals on hunting expeditions. Such response patterns indicate we have much in common with other primates. Although innately defined, in humans, those responses become more refined through learning and are exploited for cultural purposes (Sperber and Hirschfeld, 2004).

In view of the fact the human visual system has much in common with other primates, the possibility arises of being able to identify key-stimuli relevant to our distant ancestors. Alcock (1998) suggests the propensity of the human visual system to detect the outline contour of objects is predetermined as this "encourages" the detection of critical stimuli such as concealed animals; a proclivity that occurs at higher levels of discrimination and which is a universal attribute of perception (Alcock 1998). Correspondingly, Gombrich (1982) alludes to the fact we do not acquire knowledge of important aspects of the environment, such as teeth and claws, the same way as say aircraft. And, similarly emphasises that human survival depended on the recognition of meaningful features best avoided that humans are predisposed to detect. Therefore, the greater the biological relevance of a feature, the easier it will be to detect. This underlines the necessity of a pre-determined starting point without which it would be difficult, if not impossible, to acquire an ability. In effect, we see that some responses are easily triggered, whereas others require greater learning depending on the significance of particular items during evolution.

Taking all the above into account, it is clear that the essential factor to which I wish to draw attention is "sign-stimuli". This is because the typical-viewing-outline of animals encapsulates all the necessary key features for detecting animals by exploiting minimum input with maximum effect. Thus, the salient contour of the spinal column, neck, head and antlers/horns/ served as sign-stimuli crucial to human survival, and which provided an effective template that could be augmented at higher levels of the visual hierarchy; a template that was eventually co-opted for making figurative depictions of animals.

At this point, it is opportune to remind the reader that the curved sideways contour suggestive of animals is universal in palaeoart, which parallels the importance of this configuration for the visual cortex and emotional brain. As such contours were crucial to the survival of hominins over a period of three million year or more, it is not surprising they became extensive in palaeoart (Halverson, 1992; Kennedy and Silver, 1974), as Figures 2-3 and 2-4 illustrate.

Confirmation is provided by cultures lacking a figurative tradition, where individuals shown animals depicted in profile for the first time are able to recognise them almost immediately (Halverson,1992; Deregowski, 1989). This was not at first obvious because researchers tended to show animals depicted in unusual foreshortened views or with interior detail added. For example, unusual views tend to show an animal's head folded into the body. Another complicating factor was the use of texture as part of the interior detail thereby creating accidental "camouflage" making the turned animal's head even more difficult to discern. Having identified those confounds, it was concluded that the most effortless means to identify an animal in pictures across cultures is exclusively the outline profile (Halverson, 1992; Kennedy and Silver, 1974; Kennedy, 1974; Kennedy and Ross, 1975; Deregowski and McGeorge, 1998).

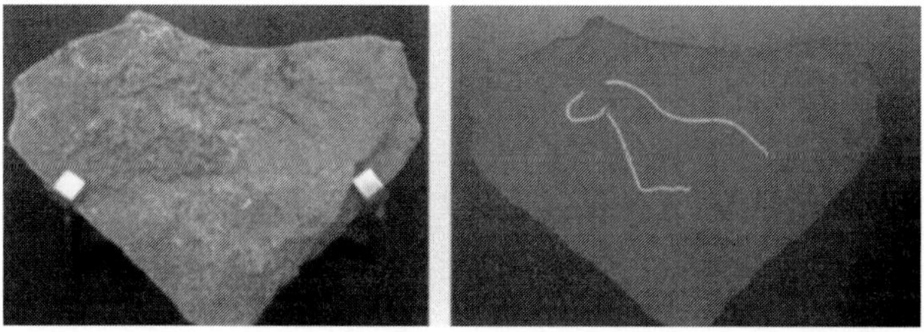

Figure 2-3. Horse engraving from Longueroche depicting cervico-dorsal contour or the "sign-stimuli" of a horse. (Photo: Don Hitchcock, 2014)

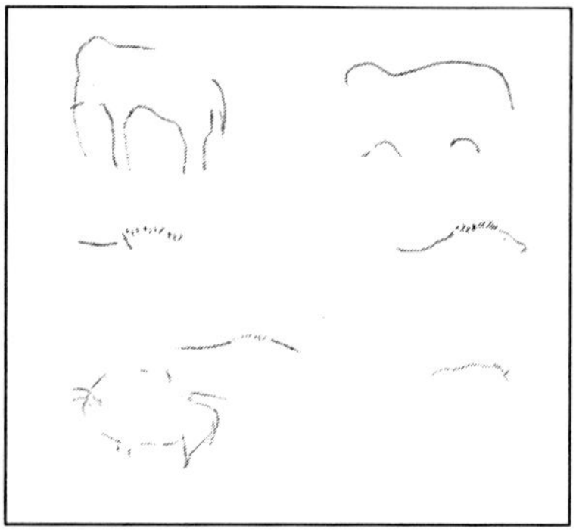

Figure 2-4. Typical examples of cervico-dorsal line as "sign-stimuli" Top: Mammoth (Cougnac Cave). Middle: Horse (Cougnac Cave). Bottom: Bison (Réseau Clastres, Ariège).

That realisation adds weight to the idea that the outline profile contour represents, in graphic form, the reification of the mental sign-stimulus for animals. In addition, those from cultures bereft of a figurative tradition tend to see particular parts of the general outline first before connecting the disjointed features together to make a coherent form, which reflects the way the visual system discerns objects when confronted with an ambiguous array (Roelfsema and Houtkamp, 2011). First, the whole field is scanned to ascertain any changes in texture that could signify an edge. Then, the edges are aligned together to encode figure from ground to define lines that indicate shape, after which each feature is linked to create the basic outline template. Any one feature or array of features can trigger this template in a way that facilitates identification to signify the whole (Tversky and Hemenway, 1984).

That capacity is borne out by the fact that the visual system evolved to defeat the camouflage and deceptive ploys of predators and prey populating the same ecological niche as hominins (Troscianko et al. 2009). It is no coincidence the visual arrays employed by psychologists for determining the principles underlying perception (Treisman and Gelade, 1980; Wolfe, 2003), such as the famous ambiguous spotted Dalmatian, are similar to the camouflage arrays with which palaeo-humans had to contend in order to detect fauna. The way observers begin to detect animals in such cases also begins with finding distinctive features that are eventually joined together to make the complete animal by filling-in the missing parts with top-down information from the "mind's eye". The fact that evolution furnished animals with camouflage, which masks their defining contours while lurking in undergrowth, speaks volumes about how the human visual system evolved. Namely, that there was an intimate reciprocal engagement between the strategies employed by animals to remain concealed and the ability of the hominin visual brain to counteract such strategies on hunting forays.

Although the profile view of animals in palaeoart is often embellished by shading, interior details, or colour, these seem to reflect higher-order recognition procedures that provide explicit confirmation as to the identity of an object. It is widely accepted that the interior details and colour play a subsidiary role to the outline contour of the overall form (Biederman and Ju, 1988). Not least because the colour of an animal is unreliable for the purpose of identification as this tends to change according to season. That said, there are further reasons to suppose the above analysis is not far from the truth.

Despite the fact evolution has pared down the perception of animals to the most rudimentary essential features, exaggerating such features in palaeo-depictions had the effect of amplifying the frequency and intensity of the intended response. This is achieved by increasing the size of the releasing factor, which then becomes a super-stimulus, which is reflected in the repeated exaggeration of certain key features in palaeoart (Mithen, 1998; Bahn and Vertut, 1997; Cheyne et al. 2009). This accords with the notion of a "supernormal stimuli" that suggests exaggeration provided a means of verifying and authenticating fundamental perceptual cues that preferentially activate the recognition system.

Mimicry in nature also employs exaggeration to trigger particular behaviours (we will investigate mimicry in detail in a later chapter). For example, by emphasising certain aspects of an attribute, such as colour, form, or pattern, such traits become more visible at a distance. Organisms eliciting such cues may therefore be using supernormal stimuli to maximally exploit the behavioural tendencies of another species. As Hinde (1982) points out, the most effective stimulus is not necessarily the one resembling the natural stimulus, but those which exaggerate obvious aspects of a natural releaser that generally lead to a more intense response. Think of the cuckoo that uses the colouring of the eggs of a much smaller bird to dupe the host into believing the larger cuckoo egg belongs to the host's clutch.

Ramachandran and Hirstein (1999) show that exaggeration in art and super-stimuli are similarly disposed by referring to the fact that, when an animal is conditioned to discriminate between a square and a rectangle and is rewarded for choosing the rectangle, it very quickly learns to respond with greater rapidity to the rectangle. Crucially, when presented with a rectangle even more rectangular in shape (thinner) the response is even greater. As such exaggeration is common in palaeoart, it seems obvious that the key-stimuli referred to were accentuated in order to create super sign-stimuli. Palaeo-artists may therefore have been accentuating particular anatomical parts in order to denote their importance for detecting and identifying animals on hunting forays.

## Notes

1. Going from the periphery of the human eye to the central fovea is akin to moving back in evolutionary time (Gregory, 1966). This trajectory can be applied to the pathway leading from the retina to higher stages of the brain. Thus, it may be that as we move down from the higher visual brain through each level we are also going back in evolutionary time, a scenario that applies to other higher sensory brain systems (Mashour and Alkire, 2013). This is probable because the eye initially developed from a simple mechanism to process changes in light, such as night and day, or shadows. Similarly, the human eye, if it were to be partitioned from the visual brain, would, computationally, only process relatively crude information. The evolution of the eye was therefore built on previous innovations that promoted survival that gradually accumulated to become more sensitive and specific to things in the world. In essence, from the eye to the visual brain, we witness a history of the levels of processing that occurred over millions of years. The result is that, as we move up from the eye to the higher visual cortex, responses become more sophisticated with links to the emotional brain modulating behavior. Such implicit processes consist of "simple" basic cues where there is less opportunity to countermand their urgency (therefore more automatic). The higher levels of the visual brain, in responding to more elaborate kinds of information, are able to confirm or override earlier signals while, at the same time, being much more rigorous and flexible. Of course, what complicates this picture is the fact that, in the human brain, there are extensive feedback loops that modulate early predisposing criteria.

## References

Aiello, L. C. and Wheeler, P. 1995. The Expensive-Tissue Hypothesis: The Brain and the Digestive System in Human and Primate Evolution. *Current Anthropology*. 36(2): 199-221.

Alcock, J. 1998. *Animal Behaviour: An Evolutionary Approach*. Sunderland, MA:Sinauer Associates.

Altman, M. N., Khislavsky, A. L., Coverdale, M. J. and Gilger, J. W. 2016. Adaptive attention: how preference for animacy impacts change detection. *Evolution and Human Behavior*. 37: 303–314.

Bahn P. G. and Vertut, J. 1997. *Journey Through the Ice Age*. London: Weidenfeld and Nicolson.

Biederman, I. and Ju, G. 1988. Surface versus edge-based determinants of visual recognition. *Cognitive Psychology.* 20: 38-64.
Brain, C. K.1981. *The Hunter or the Hunted.* Chicago: University of Chicago Press.
Cahill, L., Haier, R. J., Fallons, J., Alkirke, M. T., Tang, C., Keator, D., Wu, J. and McGaugh, J. L. 1996. Amygdala activity at encoding correlated with long-term, free recall of emotional information. *Proceedings of the National Academy of Sciences USA (Neurobiology).* 93: 8016-8021.
Caramazza, A. and Mahon, B. Z. 2003. The organization of conceptual knowledge: the evidence from category-specific semantic deficits. *Trends in Cognitive Sciences.* 7(8): 354-361.
Cheyne, J. A., Meschino, L., Smile, D. 2009. Caricature and contrast in the Upper Palaeolithic: Morphometric evidence from cave art. *Perception.* 38: 100-108.
DeCasien, A. R., Williams, S. A. and Higham, J. P. 2017. Primate brain size is predicted by diet but not sociality. *Nature: Ecology and Evolution.* 1(0112). doi: 10.1038/s41559-017-0112
Deregowski, J. B. 1989. Real space and represented space; Cross-cultural perspectives. *Behavioral and Brain Sciences.* 12: 51-119.
Deregowski, J. B. and McGeorge, P. 1998. The role of typical contours in the encodement of objects in 3-D arrays. *Perception.* 27: 283-294.
Dziurawiec, S. and Deregowski, J. B. 2002. The eyes have it: a perceptual investigation of eyespots. *Perception.* 31:1313-1322.
Folk, C. L. and Remington, R. 2006. Top-down modulation of preattentive processing: Testing the recovery account of contingent capture. *Visual Cognition.* 14:445-465.
Gardner, B. F. and Wallach, L. S. 1965. Shapes of figures identified as a baby's head. *Perceptual and Motor Skills.* 20: 135-142.
Gilbert C. D. 1996. Plasticity in visual perception and physiology. *Current Opinion in Neurobiology*. 6: 269–274.
Gombrich, E. H. 1982. *The Image & The Eye - Further studies in the psychology of pictorial representation.* London: Phaidon.
Gomes, N., Silva, S., Silva, C. F. and Soares, S. C. 2017. Beware the serpent: the advantage of ecologically-relevant stimuli in accessing visual awareness. *Evolution and Human Behavior.* 38(2): 227-234.
Gregory, R. L. 1966. *Eye and Brain.* London: Weidenfeld and Nicolson.
Halverson, J, 1992. The first pictures: perceptual foundations of Paleolithic art. *Perception.* 21: 389-404.
Hart, D. and Sussman, R. W. 2009. *Man the Hunted: Primates, Predators, and Human Evolution*. Boulder, CO: Westview Press.

Hinde. R. A. 1982. *Ethology*. London: Fontana.

Hodgson, D. 2004. Ways of Seeing: The Innocent Eye, Individual View and Visual Realism in Art. *Journal of Consciousness Studies*. 11(12): 3-16.

Isbell, L. A. 2006. Snakes as agents of evolutionary change in primate brain. 2006. *Journal of Human Evolution*. 51: 1-35.

Kennedy, J. M. 1974. Is gradual recognition of pictures by Ethiopian subjects a differentiation process or a pattern construction process? *Perception*. 3: 29-31.

Kennedy, J. M. and Ross, A. S. 1975. Outline picture perception by the Songe of Papua. *Perception*. 4: 391-406.

Kennedy, J. M. and Silver, J. 1974. The surrogate functions of lines in visual perception. Evidence from antipodal rock and cave artwork sources. *Perception*. 3: 313-312.

Koch, C. and Crick, F. 2001. The Zombie Within. *Nature*. 411: 893.

Lea, S. E. G. 1984. *Instinct, Environment and Behaviour*. London: Methuen.

LeDoux, J. E. 1994. Emotion, Memory and the Brain. *Scientific American* (June):32-39.

—. 1998. *The Emotional Brain*. London: Weidenfeld and Nicolson.

Li, F. F., Vanrullen, R., Koch. C. and Perona, P. 2002. Rapid natural scene categorization in the near absence of attention. *Proceedings of the Natonal Academy of Sciences USA*. 20: 9596-9601.

Lorenz, K. and Tinbergen, N. 1938. Taxis und Instinkthandlung in der Eirollbewegung der Graugaus. *Zeitsdrift für Tierpsychologie*. 2: 1-29.

Martin, A. 2007. The Representation of Object Concepts in the Brain. *Annual Review of Psychology*. 58: 25-45.

Masataka. N., Hayakawa, S. and Kawai, N. 2010. Human Young Children as well as Adults Demonstrate 'Superior' Rapid Snake Detection When Typical Striking Posture is Displayed by the Snake. *PLoSONE*. 5(11): e15122. doi:10.1371/journal.pone.0015122

Mashour, G. A. and Alkire, M.T. 2013. Evolution of consciousness: Phylogeny, ontogeny, and emergence from general anesthesia. *Proceedings of the National Academy of Sciences USA*. 110 (2):10357-10364.

Mithen, S. J. 1988. Looking and Learning: Upper Palaeolithic Art and Information Gathering. *World Archaeology*. 19 (3):297-323.

Munar, E., Gómez-Puerto, G., Call, J., Nadal, M. 2015. Common Visual Preference for Curved Contours in Humans and Great Apes. *PLoSONE*. 10 (11): e0141106. doi:10.1371/journal.pone.0141106

Murphy, S. T. and Zajonc, R. B. 1993. Affect, cognition, and awareness: Affective priming with optimal and suboptimal stimulus exposures. *Journal of Personality and Social Psychology*. 64(5): 723-739.

New, J., Cosmides, L. and Tooby, J. 2007. Category-specific attention for animals reflects ancestral priorities, not expertise. *Proceedings of the National Association of Sciences USA*. 104 (42): 16598-16603.

New, J. J. and German, T.C. 2015. Spiders at the Cocktail Party: An Ancestral Threat that Surmounts Inattentional Blindness. *Evolution and Human Behavior*. 36(3): 165–173.

Öhman, A. and Soares, J. J. F. 1998. Emotional Conditioning to Masked Stimuli: Expectancies for Aversive Outcomes Following Nonrecognised Fear-Relevant Stimuli. *Journal of Experimental Psychology: General*. 127 (1): 69-82.

Pessoa, L and Adolphs, R. 2010. Emotion processing and the amygdala: from a 'low road' to 'many roads' of evaluating biological significance. *Nature Reviews Neuroscience*. 11(11): 773–78.

Ramachandran, V. S. and Hirstein, W. 1999. The Science of Art - A Neurological Theory of Aesthetic Experience. *Journal of Consciousness Studies*. 6 (6/7): 15-51.

Roelfsema, P. R. and Houtkamp, R. 2011. Incremental grouping of image elements in vision. *Attention, Perception & Psychophysics*. 73(8): 2542-72.

Schmidt, F. and Vankleef, K. 2016. Response priming evidence for feedforward processing of snake contours but not of ladder contours and textures. *Vision Research*. 126: 174–182.

Seyfarth, R. M., Cheney, D. L. and Marler, P. 1980. Monkey responses to three different alarm calls: evidence for predator classification and semantic communication. *Science*. 210: 801-803.

Shepard R. N. 1984. Ecological Constraints on Internal Representation: Resonant Kinematics of Perceiving, Imagining, Thinking, and Dreaming. *Psychological Review*. 91 (4): 417-447.

Sperber, D. and Hirschfeld, L. A. 2004. The cognitive foundations of cultural stability and diversity. *Trends in Cognitive Sciences*. 8(1): 40-46.

Tinbergen, N. 1951. *The Study of Instinct*. Oxford: Oxford University Press.

Tinbergen, N. and Perdeck, A. C. 1950. On the stimulus situation releasing the begging response in the newly hatched Herring Gull chick. *Behaviour*. 3: 1-38.

Tooby, J. and Cosmides, L. 2005. Conceptual foundations of evolutionary psychology. In, D. Buss (Ed.), *The Handbook of Evolutionary Psychology* (pp. 5–67). Hoboken, NJ: John Wiley & Sons Inc.

Treisman, A. and Gelade, G. 1980. A feature-integration theory of attention. *Cognitive Psychology*. 12: 97-136.

Troscianko, T., Benton, C. P., Lovell, P. G., Tolhurst, D. J. and Pizlom, Z, 2009. Camouflage and visual perception. *Philosophical Transactions of the Royal Society B*. 364(1516): 449–461.

Tversky, B. and Hemenway, K. 1984. Objects, Parts, and Categories. *Journal of Experimental Psychology: General*. 113 (2): 169-193.

Van de Cruys, S. and Wagemans, J. 2011. Putting reward in art: A tentative prediction error account of visual art. *i-Perception*. 2: 1035–1062.

Vida, M. D. and Behrmann, M. 2017. Subcortical Facilitation of Behavioral Responses to Threat. *Scientific Reports*. 7(1):13087.

Wolfe, J. M. 2003. Moving towards solutions to some enduring controversies in visual search. *Trends in Cognitive Sciences*. 7 (2): 70-76.

Yang, J.J., Francis, N., Bellgowan, P.S.F. and Martin, A. 2005. Object concepts and the human amygdala: enhanced activity for identifying animals independent of in-put modality and stimulus format. *Paper presented at 12th Annual Meeting of the Cognitive Neuroscience Society*. New York.

Zhu, W., Drewes, J., Peatfield, N.A. and Melcher, D. 2016. Differential Visual Processing of Animal Images, with and without Conscious Awareness. *Frontiers in Human Neuroscience*. 10(513): 1-19.

# CHAPTER THREE

# CATEGORIES, CUES, AND PROFILES

> Recognition from outline shape probably requires that the object concerned is both well known and has a particularly characteristic shape, and will often also require that it is seen (or depicted) from one of a limited range of viewing positions. (Ellis and Young, 2013, p. 28)

## Diagnostic cues

On hunting expeditions, the first hint as to the identity of a distant or camouflaged animal consists of the uppermost anatomical features. For example, modern-day hunters pay particular attention to diagnostic cues (Mithen, 1988) such as antlers, horns, head or upper spinal profile (known as the cervico-dorsal line). We should therefore expect those cues to be well represented in UP art. This is exactly what we find where such features are portrayed along the sideways contour to the extent they constitute the most commonly defined attributes. For deer, this might be the line of the back, neck, antlers and head. For aurochs, (the ancestor of domestic cattle) this is again the back, neck and head but also includes the horns. This tendency, however, is not confined to UP art as it can be found in rock art throughout the world. From a practical point of view, it is obvious why this occurs, the upper features remain prominent most of the time. That is, the upper parts are most likely to remain visible in long grass and, compared to lower features, stand out at a distance, particularly in the case of antlers and horns. Traits of this kind are salient because they are illuminated by light issuing from above, whereas the lower regions tend either to be in shadow or obscured by intervening objects (Chambers et al. 1999). Chambers et al (1999) also found that people typically attempt first to identify the upper portions of objects even when presented with random, unnatural figures and the effect is more apparent with objects roughly the size of the human figure or smaller. This is because objects in the real world have a tendency to remain upright and have tops that are informative about functionality.

Importantly, the tops of animals almost universally contain the sensory apparatus and, as such, provide vital clues as to the behaviour of an animal. In cluttered environments where animals employ camouflage to remain concealed, as in Figure 3-1, the distinctive features invariably consist of the topmost anatomical parts. Interestingly, when an ibex is initially identified in this image, it seems to emerge from the rocks beginning with the horns, head and spinal contour in a similar way that

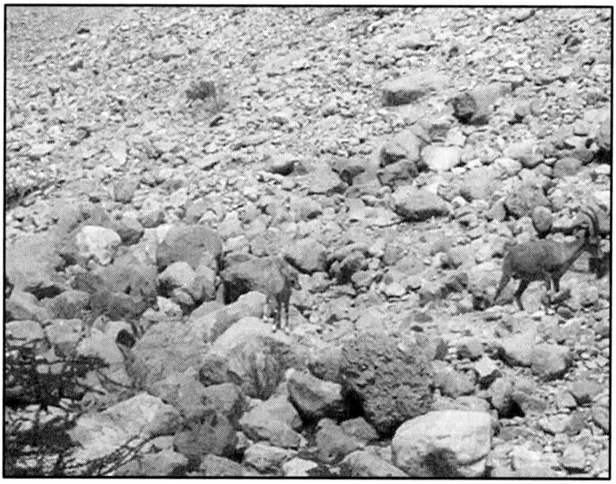

Figure 3-1. Three Ibex in rocky desert environment. (Creative Commons Attribution-Share Alike 3.0 Unported license)

cultures without a figurative tradition first begin to detect animals in pictures as outlined in the previous chapter. The prominence of the upper features is matched by the lack of legs and feet in UP art, although the underbelly is often portrayed. In fact, the hooves are invariably the only part of the anatomy missing even when the complete outline is included.

The above heuristics seem to have been incorporated into the functioning of the visual brain in precise rules, which dictate not just the preeminence of the cervico-dorsal contour but also how shape from shading is perceived. One of those rules states that the visual system assumes light comes from above rather than below, which leads to certain expectations as to how 3D forms are perceived (Ramachandran,1995). This may be one reason many animals have counter-shading in the sense that tops tend to be darker and bottoms lighter in tone, known as Thayer's law. The importance of the top-most features in palaeoart is reinforced by the fact that—even when the full outline is drawn—the upper lines are nearly always depicted first, beginning with the head, spinal contour and upper neck, or alternatively, followed by the head and front part of the neck/breast and then the back (Fritz,1999; Fritz and Tosello, 2000). In contrast to secondary anatomical traits e.g., colour, coat texture etc., such diagnostic features are prioritised because they stand out from an animal's main profile and, are frequently accentuated through repeated overpainting or over-engraving of previously made contours (Fritz, 1999; Ucko and Rosenfeld,1967). Such features stand out and consequently provide an efficient and rapid means to detect animals in challenging environmental situations. Rules of this order seem to be encoded in the visual brain, which suggests the UP depiction of animals constitute an external manifestation of the internal workings of the visual cortex, a theme to which we will return in due course.

## Basic and general categories

In tandem with the typical outline view, a further strategy the visual system employs to classify animals concerns attributes that co-occur. For example, "feathers" and "wings" often co-occur but the combination of "feathers" and "wheels" do not. Clusters of such features form the basis of a natural classification of objects—that is, on seeing one attribute it is likely certain other traits will also be apparent.

Another strategy performed at the preliminary classification stage concerns registering generalised features at the superordinate level, the most wide-ranging category, such as “animal”, “vehicle”, “plant” etc. The next stage of increasing detail is the basic level and is usually the most useful for ongoing purposes e.g. cat, dog, horse, sheep, which has high cue validity. In other words, this level contains the greatest number of inclusive features for identifying most category members. For example, the cue "long neck" has high cue validity for the category giraffe because it predicts giraffes (Logothetis and Sheinberg, 1996; Rosch et al. 1976). A further cue for giraffes might be exceptionally long legs. Cue validity is defined as the sum of the defining features valid for each category of animal. This is usually referred to as the entry-level stage of recognition because it springs to mind with ease when one needs to categorise an item. The final level is the subordinate stage, where a more detailed representation of a category is encoded; for dogs this would be Bulldog, Pekinese, Greyhound, Great Dane etc., which may well involve supplementary details such as colour, texture, coat, interior highlights etc.

It was originally thought the basic level represents the earliest and fastest to be accessed. Recent research, however, suggests that, in order to ascertain the general object class in the first instance, e.g. whether it is actually an animal, the visual system initially accesses the superordinate level (Vanrullen and Thorpe, 2001). This level seems to be based on coarse features and tends to be tagged before more detailed levels, especially when rapidity and brief exposure are of the essence (Mack and Palmeri, 2015). The fact that animals are made up of particular kinds of curved contours is crucial here, especially as they appear to be encoded pre-consciously. This parallels the same kinds of lines found in UP art, namely the many short curved lines that suggest animal forms, as illustrated in Figures 2-3 and 2-4 above. Deregowski (Deregowski and McGeorge, 1998; Deregowski et al. 1996) has shown how the degree to which a line is curved (its "pronouncedness") is crucial in discriminating objects (see also Attneave, 1954), which, as specified earlier, is especially relevant to animal forms. In fact, the entire range of categorical features exist in UP art from the superordinate level, represented by single curved contours, to the basic level, indicated by the outline contour and salient features—such as neck and dorsal outline (the most common)—to the subordinate level specified by the interior details and colour of individual animals (Halverson, 1991).

The progression from the superordinate to basic to subordinate level is exactly what we would expect considering the significance of animals to early hunter-gatherers. That is, they were concerned with cues initially identifying an animal as an animal at the superordinate level, then the species of animal at the basic level (the most important category), but somewhat less concerned with the details of individual animals at the subordinate level. A passing interest in the subordinate level is supported by the fact that some depictions focus on subsidiary features, such as sex, age, or distinguishing marks, which suggests some palaeo-artists sought to improve their recognition skills beyond the basic level. The subordinate level, however, is often regarded as more widespread than is actually the case due to the fact that the famous caves such as Lascaux, Altamira and Chauvet are most often foregrounded, whereas, for the majority of caves, abbreviated animal outlines predominate. Even in the celebrated caves a significant number of depictions consist of simple outline contours. From this, we can surmise that hunter-gatherers *occasionally* augmented the more basic visual cues for some animals in UP depictions.

To give a modern example, farmers are able to identify individual cattle or sheep whereas the layman is limited to identifying animals at the generalised basic level (Logothetis and Sheinberg, 1996; Tanaka, 2001). In other words, for most people all sheep tend to look the same. This phenomenon can be observed in bird watchers who, as experts, can distinguish subtle differences within bird taxa whereas, for naïve observers, the same birds remain indistinguishable. For experts, the subordinate level has become one of greater refinement (and therefore becomes a new basic level). Thus, certain objects in particular situations are liable to improvement through visual feedback. Experts, who spend inordinate time identifying things such as birds, are able to achieve this because they engage in explicit feedback routines that fine tune basic visual abilities (Tarr and Cheng, 2003). In general, palaeo-artists seemed content to rely on the basic-level of lion, bear, horse, because that level adequately met their everyday needs for the identification of a particular species for hunting purposes. What the archaeological record therefore tells us is that palaeo-artists largely chose to depict the salient profile outline at species level and chose to avoid depicting individual animals. Even so, we need to keep in mind that palaeo-artists endeavoured to make the recognition of such views challenging and it is to how this challenge was achieved to which we now turn.

## All in the mind's eye?

The perceptual/recognition system rationalizes visual information in order to facilitate efficient seeing by exploiting information in sets of typical (primary) views, or usual views. These views provide a shorthand means of dealing with the visual world to enable the categorization of objects with minimum effort, hence facilitating an effective, practical interaction with the everyday. This is especially useful when animals need to be identified in the wild, but how is this translated into depicting animals from visual memory? Recent research on the "mind's eye" shows that the image in visual memory tends to be canonical i.e., the most familiar view, which, in the case of animals, is the typical sideways view (Blanz et al. 1999). As indicated, the image in visual memory is derived from immediate recognition (Klein et al. 2001), which is based on the principle that the maximum amount of information needs to be encoded with minimum "effort" (Roth and Bruce, 1995). That principle is reinforced by the fact that, when asked to state a preference from a selection of different views of a known object, we tend to choose the usual rather than unusual view (Palmer et al. 1981).

Obviously, an artist is, to some extent, dependent on visual memory, as, even when the object to be depicted is immediately available, one has still to summon up an image of the scene when actually putting pencil to paper. The important point here is that imagery is a kind of by-product of the perceptual system and, accordingly, depends on the same fundamentals.

By understanding those procedures, we are better placed to appreciate why palaeo-artists first began to ply their trade.

The preceding insights go a long way to explain how the typical-viewing-profile came to predominate in UP art and answers the question of why palaeo-artists neglected unusual or foreshortened views of animals in depictions. Namely, as the typical viewing profile embodies fundamental aspects of visual information for encoding form, what Deregowski (2017) refers to as "significant contours", the need to portray unusual views did not arise. Thus, we should expect to find only a modicum of frontal views of animals in palaeo-depictions (the shorter, minor axis or subsidiary view). In fact, frontal-views constitute the second most frequent angle of view e.g., the four (perhaps five) bison heads of Chauvet (Clottes, 2003), as well as lion, ibex and horses from the Magdalenian period; the frontal-views found in mobiliary art such as the engraved horse from Las Caldas, Asturias (Bahn and Vertut, 1997: pp.134, 135), La Madeleine (Paillet, 1996: Figure 2), and Gourdan (Graziosi, 1956: Plate 88). The evidence from palaeoart therefore confirms what neuroscience and perceptual psychology predicted. As visual science discovered how the visual system works independently of palaeo-artists, this reinforces the importance of the above insights and brings new and crucial knowledge to bear on palaeoart. Understanding how perception and recognition proceeds, hence, provides a better appreciation as to why animals were portrayed in particular ways during the UP. In short, we are able to peer into the minds of hunter-gatherers, which suggests they depended on and were in competition with particular animals on a number of levels. This was a codependence that led to the need for an efficient means of identifying fauna exemplified by the sideways view that was coopted to depict animals. Paradoxically, the carnivores that were in competition with palaeo-humans and the prey they hunted probably helped configure this mental prototype in that those with the more efficient visual system stood a better chance of survival. Through a gradual ratcheting-up, the visual brain became more proficient at detecting animals. In that sense, the predators that feasted on palaeo-humans, and the prey on which humans depended for protein gave rise to —not just the way they were configured in the visual brain—but ultimately how they were depicted in palaeoart. Ironically, one might speculate that animals came to paint themselves through us!

Those aspects of the portrayal of animals relate to how the brain functions at mid-levels, but as the neural substrates for detecting animals go deep into our evolutionary past, there are further clues to be gained from the visual brain and psychology of perception.

## References

Attneave, F. 1954. Some informational aspects of visual perception. *Psychological Review*. 61: 183-193.

Bahn P. G. and Vertut, J. 1997. *Journey Through The Ice Age*. London: Weidenfeld and Nicolson.

Blanz, V., Tarr, M. J. and Bülthoff, H. H. 1999. What object attributes determine canonical views? *Perception*. 28:575-599.

Chambers, K. W., Mcbeath, M. K., Schiano, D. J. and Metz, E. G. 1999. Tops are more salient than bottoms. *Perception and Psychophysics*. 61 (4): 625-635.

Clottes, J. 2003. *Return to Chauvet*. London: Thames and Hudson.

Deręgowski, J.B. 2017. The Psychology of Graphic Perception. In, B. David and I. J. McNiven (Eds.), *The Oxford Handbook of the Archaeology and Anthropology of Rock Art, (Oxford Handbooks Online)*. Oxford: Oxford University Press.

Deregowski, J. B. and McGeorge, P. 1998. The role of typical contours in the encodement of objects in 3-D arrays. *Perception*. 27: 283-294.

Deregowski, J. B., Parker, D. M. and Dziurawiec, S. 1996. The role of typical contours in object processing by children. *British Journal of Developmental Psychology*. 14:425-440.

Ellis, A. W. and Young. A. W. 2013. *Human Cognitive Neuropsychology: A Textbook with Readings*. Hove: Psychology Press.

Fritz, C. 1999. Towards the reconstruction of Magdalenian artistic techniques: the contribution of microscopic analysis of mobiliary art. *Cambridge Archaeological Journal*. 9:189-208.

Fritz C. and Tosello, G. 2000. Observations on the techniques in the horse panel at the Chauvet Cave (Ardèche). *International Newsletter on Rock Art*. 26: 23–29.

Graziosi, P. 1956. *Die Kunst der Altsteinzeit*. Stuttgart: Firenze.

Halverson, J. 1991. Paleolithic Art and Cognition. *The Journal of Psychology*. 126 (3): 221-236.

Klein, I., Paradis, A. L., Poline, J. B., Kosslyn, S. M. and Bihan, D. L. 2001. Transient Activity in the Human Calcarine Cortex During Visual-Mental Imagery: An Event-Related fMRI Study. *Journal of Cognitive Neuroscience*. 12 (2): 15-23.

Logothetis, N. K. and Sheinberg, D. L. 1996. Visual Object Recognition. *Annual Review of Neuroscience*. 19: 577-621.

Mack, M.L. and Palmeri, T.J. 2015. The Dynamics of Categorization: Unraveling Rapid Categorization. *Journal of Experimental Psychology. General*.144(3): 551–569.

Mithen, S. J. 1988. Looking and Learning: Upper Palaeolithic Art and Information Gathering. *World Archaeology.* 19 (3): 297-323.
Paillet, P. 1996. A propos de trois bisons Magdaléniens exceptionels (La Madeleine, Dordogne). *Paléo.* 8: 359-366.
Palmer, S., Rosch, E. and Chase, P. 1981. Canonical perspective and the perception of objects. In, J, Long. and A. Baddeley (Eds.), *Attention and Performance* Vol. IX. (pp. 135-15). Hillsdale, NJ: Lawrence Erlbaum Associates.
Ramachandran, V. S. 1995. 2-D or not 2-D that is the question. In, R. Gregory., J. Harris., P. Heard. and D. Rose (Eds.), *The Artful Eye* (pp. 249-267). Oxford: Oxford University Press.
Rosch, E., Mervis, C. B., Gray, W. D., Johnson, D. M. and Boyes-Braem, P. 1976. Basic objects in natural categories. *Cognitive Psychology.* 8:382-439.
Roth, I. and Bruce, V. 1995. *Perception and Representation - Current Issues.* Buckingham: Open University Press.
Tanaka, J. 2001. Hierarchies, similarity, and interactivity in object recognition: "Category-specific" neuropsychological deficits. *Behavioral and Brian Sciences.* 24:493-494.
Tarr, M. J. and Cheng, Y. D. 2003. Learning to see faces and objects. *Trends in Cognitive Sciences.* 7 (1): 23-30.
Ucko, P. and Rosenfeld, A. 1967. *Palaeolithic cave art.* London: Weidenfeld and Nicolson.
Vanrullen, R. and Thorpe, S. J. 2001. Is it a bird? Is it a plane? Ultra-rapid categorisation of natural and artifactual objects. *Perception.* 30: 655-668.

# CHAPTER FOUR

# THE EFFICIENT VISUAL BRAIN

> It is possible, of course, that Palaeolithic artists used the lighting of engravings to their advantage, making them appear and disappear to great effect—perhaps even for fun! (Bahn, 2016, p.167).

## The "expedient" and "discerning" visual brain

One of the surprising aspects of the brain is that each side deals with visual information differently. It is commonly thought that the left side is the more rational and conceptual whereas the right tends to be artistic and creative. Not only have these differences been exaggerated but ignore the subtle discrepancies in how vision is assimilated by each hemisphere. Even at lower levels there may be differences in discriminating detail and global form according to attentional demands (Dereulle and Fagot, 1997; Fink et al. 1997), though others question this (Sasaki et al. 2001). At higher levels of the association cortex, this tendency becomes increasingly salient. Note, however, the division between hemispheres is one of degree or bias rather than absolute (Mazard et al. 2005). The left hemisphere appears to be concerned with generalization in its ability to recognise an object in a number of different scenarios (achieve constancy) thanks to the limited number of recognition templates that contain what I refer to as "foremost primary" and "foremost secondary" views, which expedite familiarisation (see Figure 4-1). This hemisphere is recruited each time a known object is encountered and provides a fast-track route to recognition (referred to as the "expedient eye"), which interfaces with language at higher stages of the same hemisphere. The left hemisphere has accordingly been named "The Interpreter" by Michael Gazziniga (Gazzaniga et al. 2014). On the savannah, hominins would have engaged this side of the brain in most situations to identify animals. In other words, the left hemisphere came into play when a hunter saw an animal from the usual (foremost primary view) and the more accessible unusual perspectives (foremost secondary views) as illustrated in Figure 4-1. As speed is of the essence, it is likely the left hemisphere—in conjunction with the aforementioned subcortical route—encoded the critical key-stimuli for perceiving animals. However, in more challenging situations where animals lay hidden in camouflage, dim light or shadows, the right hemisphere with its wider "spotlight" takes over (Ramachandran and Blakeslee, 1998). Such a capacity allows the environment to be scanned panoramically for any suspicious looking entities. As such, the right hemisphere can alert and bring to focal attention any suspicious looking objects worthy of scrutiny by the left visual brain. Thus the right hemisphere, is more sensitive to differences in overall pattern and texture across a wider viewing angle. Evidence therefore suggests that attention to global features employs a wide attentional spotlight whereas attention to local features depends on a narrower spotlight (Robertson et al. 1993; Sasaki et al. 2001; Weissman and Woldorff, 2005). These hemispheric asymmetries may be due to the fact right and left temporo-parietal regions are biased for relatively low frequencies (encoding global features) and relatively high spatial frequencies (encoding local features), respectively (Weissman and Woldorff, 2005). In effect, the left hemisphere is particularly astute in dealing with detail, a propensity that is useful for focussing on an object to determine its identity.

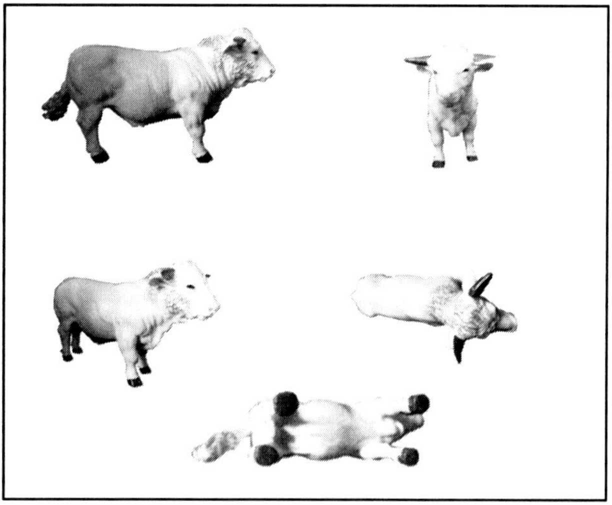

Figure 4-1. Different perspective views of a quadruped. Top left: Foremost primary view or typical viewing profile (most commonly accessed). Top right: Foremost secondary view (second most common). Middle left: Subsidiary view. Middle right: Unusual view. Bottom: Anomalous view.

The right visual brain is referred to as the "discerning eye" as it appears to act as a back-up, or additional resource in situations where the left hemisphere is unable to recognise an object in challenging situations—possibly involving camouflage, poor lighting conditions or anomalous/subsidiary views (Warrington and James, 1988; McCarthy and Warrington, 1990;Warrington and Taylor, 1978; Warrington, 2009). So, although a structural description (the "basic" features produced by lower brain areas)[1] as an output of the early visual cortex is relayed to both hemispheres, the higher right side holds the information for future reference in circumstances when the left hemisphere is unable to match the incoming description to the fast track network of the left hemisphere (see Figure 4-2). Visual information is therefore assigned to the right hemisphere to determine whether any further useful information might be gleaned from the initial "raw" description in the face of challenging anomalous views and when objects are degraded or particularly difficult to discriminate (Blanca and López-Montiel, 2009). That information may also be fed back to the early visual cortex in the event of exceptionally challenging visual circumstances (Hochstein and Ahissar, 2002). The feedback loop is repeated until recognition is achieved or, is otherwise referred to the attentional system for added scrutiny of the originating stimulus (the real external object) to gather more exhaustive information (Hochstein and Ahissar, 2002; see also Kosslyn et al. 1990 for a detailed description of these processes).[2] If after having submitted this information to the system several times with no positive outcome, the visual brain will render the unrecognised entity anomalous or a novel object outside the purview of everyday experience.

Interestingly, when one looks at an unfamiliar object, that is, one never before encountered, it is more likely to be processed by the right rather than the left hemisphere (Marsolek, 1999),[3] as it is equivalent to an anomalous view (see Figure 4-1and 4-2); a reassignment that is coordinated by more forward areas of the cortex within the frontoparietal system (McMenamin et al. 2016) that "decides" the hemisphere best suited to deal with perceptual information according to the current level of difficulty (Figure 4-2). Hence, idiosyncratic/anomalous views remain inaccessible to the left hemisphere because the many off-axis perspectives have not previously been assimilated by the expedient brain through successive rotations of an object in order to assign typical views. Interaction with unfamiliar objects through experience with different angles is therefore essential before neural coding and familiarisation can take place. Depending on the number of views experienced, some views will become diagnostic and consigned to the left hemisphere ("expedient"

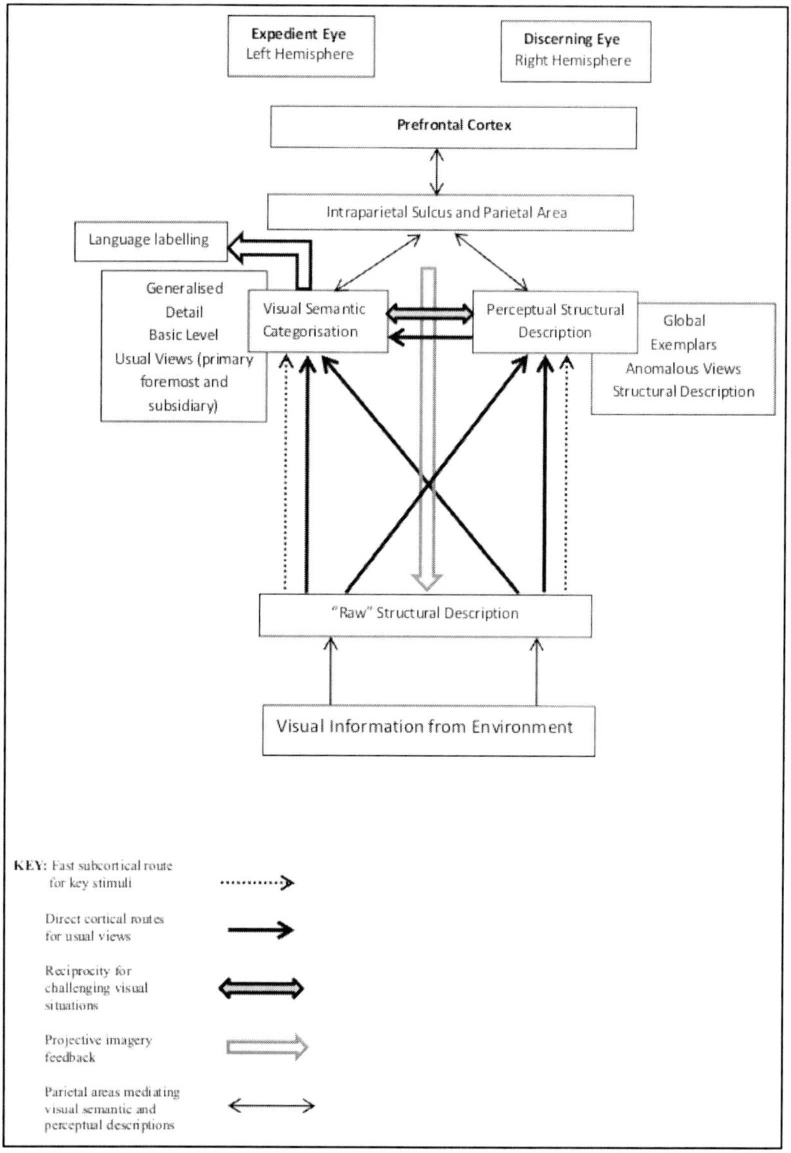


Figure 4-2. Neural routes of the "Expedient" and "Discerning" eye based on models discussed in the text and view-dependent descriptions of visual recognition (Peissig and Tarr, 2007).[4]

eye") ultimately becoming familiar through the instilling of "first order neurons". For example, if one were first to see only four separate views of a novel object, on seeing the same views again they are likely to be treated as four separate objects. However, if one initially sees, say, ten different views of the same object the visual system will have received enough information to apportion prototypes (typical views) that facilitates categorisation useful on subsequent occasions when the same object is encountered (Logothetis et al. 1996). The object will, therefore, subsequently be recognised as the same object from most angles of view. Objects that undergo such familiarization thereby become semantic percepts that tend to be channeled through the left hemisphere (Mazard et al. 2005).

As Perrett et al (1998) indicates, neural networks weakly activated by unusual views will take longer to provide sufficient input for neurons in the "decision apparatus" compared to view-specific neurons. For example, as objects are presented at increasingly unusual orientations the information processing load tends to increase. Moreover, not only will first order neurons be activated more strongly and widely when viewing familiar objects but will reach threshold before second-order patterns of neural activation. Similarly, Blanz et al (1999) reported that, in adults, imagery of familiar objects is accessed in memory more readily in a two-dimensional canonical form as opposed to an off-axis or three-quarter view. A corresponding result for model heads (initially presented evenly through all views) found that frontal and profile views were preferentially stored in memory (Harries et al. 1991). Note, in keeping with other objects, faces are processed initially and more rapidly in the left hemisphere before the right hemisphere undertakes a more considered assessment (Meng et al. 2012).

To summarise this rather challenging section, as Figure 4-2 makes clear, the left-sided "expedient eye" will be sufficient for encoding objects in most situations, but will need to call on reserves in more challenging circumstances—which is where the right hemisphere intervenes with its "eye" for dealing with ambiguity, anomalous situations and exemplars (the right hemisphere also specialises in subordinate information or supplementary details). Thus, the "discerning eye" is particularly useful for detecting animals that may be camouflaged, partially hidden, idiosyncratic (anomalous) or otherwise degraded for various reasons.

Although visual acuity in humans is particularly adept compared to other mammals, only being matched by higher primates (Caves et al. 2018), the fact that there are many more destinations for dealing with visual information in the human higher cortex (Orban et al. 2006) means flexibility is enhanced. This is underlined by the fact that the fovea, the most sensitive part of the retina for perceiving detail, is more widely represented in the higher visual cortex in humans compared to primates (Sereno et al. 1995). As a consequence, humans are more able to see the larger picture—the global precedence effect—and ignore distracting details, a capacity that comes on stream in five to six-year-olds but which never develops to the same extent in chimpanzees (Neiworth et al. 2006). In effect, the human visual system is biased toward perceiving global form (objects and things) at the expense of local features (Navon, 1977; Peressotti, 1991), though which part of the brain deals with this at any one time, i.e., left or right hemisphere, often depends on context (Fink et al 1997; Hübner and Volberg, 2005).

Intriguingly, in nonhuman primates that bias is reversed with detail taking precedence over global form (Dereulle and Fagot, 1997). In other words, humans tend to be less distracted by local features thereby facilitating the "filling in" or completion of the missing elements in a figure; a proclivity that is useful for discerning animals hidden in noisy environments. This is reflected in the capacity of humans to easily see and complete the missing lines of a partly drawn object; a skill chimps are unable to perform due to their failure to project the global mental image onto incomplete line drawings (Saito et al. 2014). The deficiency in chimps is reflected in their inability to see illusionary lines in, for example, the Kanizsa triangle (discussed further below). Though both hemispheres contribute to the filling-in of missing lines in a figure, the ability to summon up illusory contours by way of the right hemisphere means humans are much more proficient at coping with excessively challenging stimuli (Corballis et al. 1999; Gazzaniga et al. 2014).

As visual imagery and visual memory are thought to engage some of the same parts of the visual brain as the perception/recognition pathway, the image in the "mind's eye" will derive from the heuristics of the system in general. It seems that these mid-level areas of the visual cortex, which are underwritten by evolutionary instantiated domains—for identifying such things as human faces and animals—develop early and rapidly to the extent that they can easily be tuned to achieve greater efficacy (Mazard et al. 2005). This is backed up by recent research on artificial neural networks demonstrating that, when networks are tweaked to perceive particular objects e.g., animals (Suzuki et al. 2017), the likelihood of seeing fauna is enhanced.

## Implications for palaeoart

Taking the foregoing comments about typical views into account and how each hemisphere processes information in different ways, what might this tell us about palaeoart and those responsible for its realisation? The fact that most of the depictions of animals are in typical outline profile informs us that the authors were mainly engaging the left hemisphere. Recall, that the typical viewing profile provides the maximum information with minimum effort, so that an animal can be quickly recognised using this template the first time round, thus facilitating recognition of most angles of view. This also tells us that palaeo-artists were, like us, on average left hemisphere dominant and would have been employing that part of the cortex both for the rapid identification of animals and for making depictions. As that route ("expedient eye") employs both foremost and subsidiary (primary) views for recognising animals, we would expect to find these in palaeoart, which is exactly what prevails.

A further prediction is that, if true three-quarter views turn-up in the archaeological record, they should be relatively rare. This is because accessing such views involves cross-referencing the side view (foremost primary) and with the frontal view (foremost subsidiary), which provides the co-ordinates for dealing with foreshortened views. Because this requires more processing in terms of the load exerted on neural capacity, three-quarter views are marginally more difficult to access and therefore do not spring to mind as readily as the side and front views (Kosslyn 1994; Perrett et al. 1998). Neurophysiologically, foreshortened views can be accessed by recourse to the cross-referencing of a limited number of stored key views for a single object, which is referred to as the view-dependent approach to object perception (Perrett et al. 1994; Logothetis and Sheinberg, 1996; Peissig and Tarr, 2007). Interpolation between a limited number of such crucial views would provide access to most views of an object. For example, in the case of face recognition neurons, the full-face view would predominate as the foremost primary view and then the profile as the foremost secondary view. Foreshortened and off-axis views e.g., three-quarter views etc., (and therefore visual realism) would then be accessed by recourse to the overlapping tuning curves as combined outputs of the primary neuron clusters. Neural representation of a few characteristic views is therefore capable of recovering the full array of intermediary views. In a similar way, a vast array of colours can be perceived from the three basic colour cones in the human fovea thanks to the overlapping tuning curves.

Twisted perspective, though common in palaeoart (Figure 4-3), should not be confused with true three-quarter views, as the former is peculiarly distorted and the animal does not look quite right, while the latter is more convincing. Twisted perspective typically consists of a sideways view of an animal but the horns, antlers or ears are contorted towards the frontal view (foremost view). This phenomenon seems to be a compromise between the side view (foremost view) and the frontal view (foremost subsidiary), and can be explained by the visual system being "caught in two minds." More precisely, in the case of antlers, horns, ears, and eyes, the foremost primary view consists of the frontal view, whereas the side view of the same features is subsidiary; a reversal of an animal's trunk, neck and head whereby the side-view comes first and the frontal view second. By attempting to draw antlers, ears or horns in the context of the animal's body, a discrepancy occurs between the two competing foremost primary views encoding each part of an animal's anatomy, which is resolved by a compromise that gives rise to the obvious distortions.

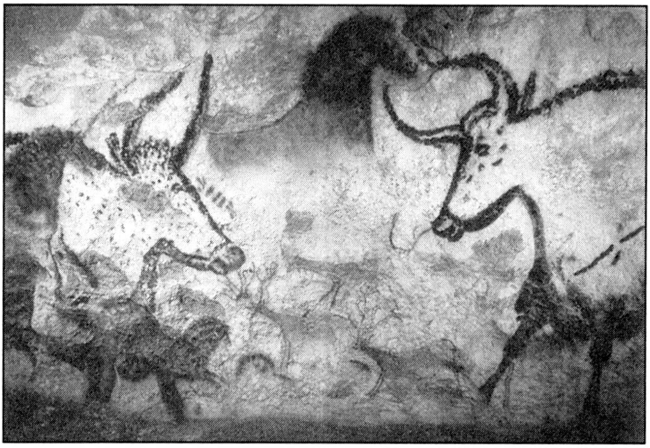

Figure 4-3. An example of twisted perspective. Note the way the horns and antlers of the animals are twisted towards the viewer. (Creative Commons Attribution-Share Alike 3.0 Unported license).

In the case of the strict side-views of animals (not in twisted perspective), the furthest legs or horns are often completely occluded by the nearer ones, so the animal appears flattened, which again reflects the visual brain's preference for foremost primary views. The above insights provide verification as to how understanding the way the visual system functions can help decipher palaeo-depictions.

Twisted perspective also relates to the fact that when we draw an object, such as a plate or wheel presented at an angle, we tend to represent the perceived ellipse as more circular than it appears. Thouless (1931) was perhaps the first to systematically investigate the phenomenon, which he labelled "regression towards the real object." In effect, we are being lured towards the foremost view by the neural mechanism outlined above. That tendency has also been observed in untrained artists and children. Later investigators, building on Thouless's insights, found that, when certain cues are controlled, even trained artists are prone to the effect (Taylor and Mitchell, 1997) to the extent the telltale signs of the tendency can be discerned in the paintings of Renaissance artists.

Those insights illustrate how form constancy is realised as a prototypical template that is overseen by the left hemisphere. The visual system thus deviously conspires to haul us back to the default mode of recognition i.e., the expeditious eye, whenever our concentration lapses or when a lack of practice intervenes. What this means is that, once the expeditious eye has labelled an object as such, attention is immediately directed to other concerns, because once an object has been labelled, and if not an immediate threat, can safely be ignored, thereby avoiding further costly processing capacity. When asked to judge the circularity of a plate or wheel we therefore tend to gravitate towards the idealised form that requires a sustained effort to counteract. And, as we saw, attending to an object from an unusual angle creates greater demands on the visual hierarchy. In one way, we could accuse the visual brain of laziness but this belittles a finely tuned apparatus that furnishes highly pertinent cues for effective interaction with the world overloaded by an unmanageable torrent of distracting information.

The depiction of hooves in palaeoart furnishes additional evidence as to the veracity of the above insights. For example, depictions of hooves appear very early in the UP, especially on stone plaques (Hodgson, 2018; Bourrillon et al. 2018) and are invariably depicted as a typical foremost view with the most obvious features portrayed. The hooves are often drawn in a similar way to an animal's hoof-print imprinted on the ground that sometimes appear as if turned towards the viewer at the base of the an animals legs, referred to as "ball feet". The portrayal of hooves may be connected to the importance of hoof-prints for stalking animals, as Guthrie (1984) and Mithen (1988,1990,1991) suggested. Mithen's analysis, in fact, goes a long way to support the notion that palaeo-depictions were closely tied to the day-to-day hunting activities of the communities involved. By depicting hooves in the way described, we can assume that they stimulated visual memory thereby helping to confirm the identity of particular animals on hunting forays. Palaeo-artists, therefore, were reinforcing the fact a particular hoof-print was associated with a certain animal so they could be more easily identified and tracked.

Another technique that achieves a similar goal concerns superimpositions where palaeo-artists deliberately created ambiguous viewing conditions. Superimpositions range from easy-to-define loose arrangements to quite dense hard-to-see configurations (see, for example, Figure 4-4). For palaeo-hunters, this kind of manipulation served to "exercise" the right hemisphere and, in tandem with the left brain, helped distinguish forms embedded in such confusing arrays.

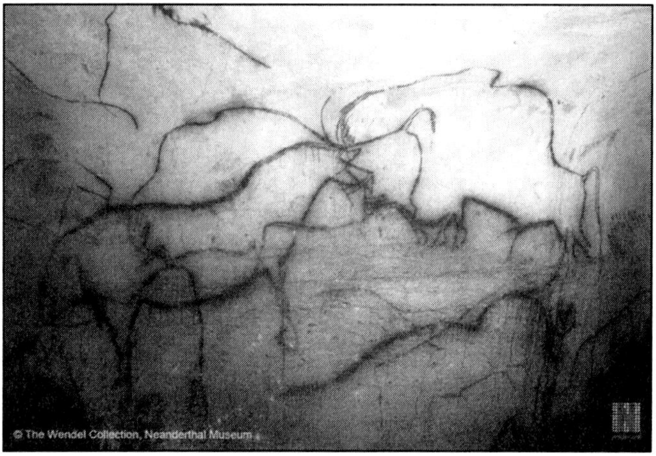


Figure 4-4. Pech Merle, Chapelle des Mammoths (© The Wendel Collection, Neanderthal Museum).

Searching for distinctive animal forms in this way facilitates the lowering of recognition thresholds, thereby improving the visual system's ability to identify animals in the real world. This is supported by the fact that a previously seen object is readily seen again when embedded in a complex array of superimposed outlines than when it is not (Kirby and Kosslyn, 1992)—mainly because visual imagery comes to the fore when one is presented with an ambiguous array of lines in that imagery tends to go into overdrive (Hegdé and Felleman, 2007). We might say that, similar to the cave environment, superimpositions acted as surrogate camouflage. As one of the main functions of camouflage is to help an animal blend into the background, the numerous overlapping lines of animals in palaeoart may have served as a means whereby a range of animals could potentially be identified in a diverse and confusing graphic array. The decorated caves are full of organic like phenomena with curious shapes and textures that simulate the outdoor natural environment of plants, trees, foliage and undergrowth. Confronted with such suggestive arrays, the visual brain of palaeo-artists entering the darkness would have been automatically activated in the ways cited. By stimulating both the left and right hemispheres through various depictive strategies, palaeo-artists were invoking the same brain dynamics that occurred when palaeo-hunters engaged in search behavior during actual hunting expeditions.

In sum, the left-hemisphere delivers an efficient means of detecting and recognising animals, which is reflected in how fauna are depicted in palaeoart. The right hemisphere, however, comes into its own in situations where objects are difficult to discern, which is an observation supported by the specialisation of the right hemisphere for dealing with remote associations (Pizzagalli et al. 2001).

## Why the typical viewing profile is typical

It is astonishing that animals in palaeoart are almost exclusively drawn from the side with the longer axis in full view (Poplin, 1984). This mode of depiction persisted despite the many angles from which an animal is viewed by widely-dispersed populations over time. As already stated, the explanation resides in the fact this view supplies the maximum amount of useful information for identifying an animal from the greatest number of directions with the minimum amount of cognitive effort and is, accordingly, programmed into the visual system. Furthermore, despite the depicted profile-view spanning at least 25,000 years, the usual explanation is that an animal is easier to depict that way (Bahn and Vertut, 1997, p.134), which does not always hold true (Deregowski and McGeorge, 1998).[5] One would have expected that the first graphic images of animals to take the form of stick-like figures, which are certainly easier to produce, yet naturalistic outlines of animals—difficult to draw in themselves—were preferred. There are other reasons why understanding how the visual brain functions can help in this regard.

We have established that most animals are depicted by recourse to the typical profile view, invariably composed of certain telling contours. (Halverson, 1992). Furthermore, the depictions are generally depicted in a static, standing posture, by far the main unit of representation. Research on how the later stages of the visual brain function supports the conclusion palaeo-artists were concerned with the single unit (Turnbull,1999; Turnbull et al. 1997), which shows the visual brain's "what" system (the "expedient eye") focusses on objects almost to the total exclusion of other factors.

As animals are invariably depicted without any reference to movement, this adds further credence to the idea that the depictions were based on perceptual factors and visual memory. There are, of course, some examples of movement portrayed and some even portray multiple limbs that may signify successive movement frames (Azéma, 2008) but both are relatively rare compared to the static depictions. The fact animals are nearly always adults substantiates the predominance of visual memory because, when we think of an animal, it usually takes the form of an adult in a stationary sideways posture (Halverson,1991). Pertinently, scenes and other types of objects such as trees, landscape features, etc., are virtually absent (Hudson, 1998; Bahn and Vertut, 1997). Clottes (2003) claims there is a herd of rhinoceroses in the "End Chamber" of Chauvet where the horns are in true perspective (converging perspective) while the bodies are in false perspective (diverging perspective). There are several reasons to suppose this is neither a scene nor in perspective. First, if the authors were aware of converging perspective, they would not produce cues directly contravening this by resorting to false perspective. Two, rhinoceroses do not stand in straight lines as represented, so this is not a view of a real scene. Three, the converging effect is probably an automatic outcome caused by the demands of repeatedly drawing curves within curves (this is supported by the fact there are similar examples in palaeoart). Four, Halverson (1992) makes a convincing case that this type of drawing should be regarded as a sequence of false starts. Deręgowski (2017) expresses similar doubts regarding the same phenomenon. From our perspective, such repetitive graphic behaviour can be regarded as a further expression of the hunter's addiction to the cervico-dorsal line and typical viewing profile that was driven by the pleasure exerted in depicting such key stimuli.

As animal outlines were accentuated by re-working, occasionally after only a short period since the creation of the original, in other cases, after a much longer interval (Bahn and Vertut, 1997), this provides compelling evidence palaeoart was about aspects of an animal's form that were crucial for their detection during hunting sorties. A raised sensitivity to animal outlines was, similarly, essential when changing encampments or in following the migration of animals.

# Notes

1. A structural description is one where incoming visual information is processed by lower-order areas of the visual cortex where lines have been grouped into coherent shapes based on figure and ground, edges and contours but have not yet been assimilated semantically (that is, meaning has not yet been assigned).

2. This scenario mirrors Richard Gregory's (1966) approach to object recognition. Gregory's constructivist approach emphasises the primacy of top-down hypothesis-testing similar to that set out here. The approach espoused in this book, however, regards hypothesis-testing as only relevant to the very later phases of visual processing when earlier stages are unable to cope with the demands exerted by incoming visual information. Although learning might be required to instill these various processing stages, the pattern by which this occurs is still underpinned by long-standing evolutionary events. Even hypothesis-testing in assessing which one of a number of competing alternatives of a given stimuli that is likely to be the correct, may ultimately be underwritten by evolutionary criteria. Hypotheses-testing is particularly relevant when a stimulus is degraded, such as in poor viewing conditions, to the extent that the visual system needs to identify the object under scrutiny quickly to determine friend or foe. In this regard, Bayesian inference is implicated as a predisposing candidate. It seems that what particularly marks out the human brain is the ability to engage in assessing the world based on prodigious forward prediction errors based on Bayesian mechanisms (Kersten and Yuilley, 2003).

3. Marsolek (Marsolek, 1995; Marsolek and Bergund, 2008) found the left cerebral hemisphere tends to classify by abstract category whereas the right hemisphere ascertains specific exemplars of an object class.

4. Viewpoint dependent approaches hold that a limited number of views of an object are encoded in a few neural clusters that are sufficient, through interpolation of such clusters, to perceive an object from most angles of view encountered in normal situations. Neural clusters become actualised through interaction with an object in an initial learning phase. This is in contrast to viewpoint independent models that posit a single structural description able to encode most, if not all, viewing angles.

5. Children sometimes draw more complex figures than the presented model calls for e.g., they tend to draw a cup with its handle showing even though the model is so oriented that the handle remains hidden (Freeman and Janikoun, 1972).

## References

Azéma M. 2008. Representation of movement in the Upper Palaeolithic: An ethological approach to the interpretation of parietal art. *Anthropozoologica.* 43 (1): 117- 154.

Bahn, P.G. 2016. *Images of the Ice Age*. Oxford:Oxford University Press.

Bahn P. G. and Vertut, J. 1997. *Journey Through The Ice Age*. London: Weidenfeld and Nicolson.

Blanca, M. J. and López-Montiel, G. 2009. Hemispheric Differences for Global and Local Processing: Effect of Stimulus Size and Sparsity. *The Spanish Journal of Psychology.* 12(1): 21-31.

Blanz, V., Tarr, M. J. and Bülthoff, H. H. 1999. What object attributes determine canonical views? *Perception*. 28: 575-599.

Bourrillon, R., White, R., Tartar, E., Chiotti, L., Mensan, R., Clark, A., Castel, J.-C., Cretin, C., Higham, T., Morala, A., Ranlett, S., Sisk, M., Devièse, T. and Comeskey, D. J. 2018. A new Aurignacian engraving from Abri Blanchard, France: Implications for understanding Aurignacian graphic expression in Western and Central Europe. *Quaternary International.* 491:46-61. doi: 10.1016/j.quaint.2016.09.063

Caves, E. M., Brandley, N. C., Johnsen, S. 2018. Visual Acuity and the Evolution of Signals. *Trends in Ecology and Evolution*. (33)5: 358-372.

Clottes, J. 2003. *Return to Chauvet.* London: Thames and Hudson.

Corballis, P. M., Fendrich, R., Shapley, R. M. and Gazzaniga, M. S. 1999. Illusory contour perception and amodal boundary completion: Evidence of a dissociation following callosotomy. *Journal of Cognitive Neuroscience.* 11:459-466.

Deręgowski, J.B. 2017. The Psychology of Graphic Perception. In, B. David and I. J. McNiven (Eds.), *The Oxford Handbook of the Archaeology and Anthropology of Rock Art.* (Oxford Handbooks Online). Oxford: Oxford University Press. doi:10.1093/oxfordhb/9780190607357.013.5

Deregowski, J. B. and McGeorge, P. 1998. The role of typical contours in the encodement of objects in 3-D arrays. *Perception.* 27:283-294.

Dereulle, C. and Fagot, J. 1997. Hemispheric Lateralisation and Global Precedence Effects in the Processing of Visual Stimuli by Human and Baboons. In, J. Fagot, L. Rogers, J. Ward, B. Bulman-Fleming and W. Hopkins (Eds.), *Hemispheric Specialisation in Humans and Animals* (pp.233-246). Hove: Psychology Press.

Fink. G. R., Marshall, J. C., Halligan, P. W., Frith, C. D., Frackowiak, R. S. J. and Dolan, R. J. 1997. Hemispheric specialization for global and local processing: the effect of stimulus category. *Proceedings of the. Royal Society of London B.* 264: 487–494.

Freeman, N. H. and Janikoun, R. 1972. Intellectual realism in children's drawing's of a familiar object with distinctive features. *Child Development.* 43: 1116-1121.

Gazzaniga, M., Richard, I. and George, M. 2014. Cognitive Neuroscience. *The Biology of the Mind*. New York: Norton.

Gregory, R. L. 1966. *Eye and Brain.* London: Weidenfeld and Nicolson.

Guthrie, R. D. 1984. Ethological observations from Palaeolithic art. In, H.G. Bandi.,W. Huber., M.R. Sauter. and B. Sitter (Eds.). *La contribution de la zoologie et de l'ethologie à l'interprétation des peuples chasseurs préhistoriques* (pp. 35-74). SSSH (3rd Colloque de la société Suisse des Sciences Humaines.) Fribourg: Editions Universitaires Fribourg.

Halverson, J. 1991. Paleolithic Art and Cognition. *The Journal of Psychology.* 126 (3):221-236.

—. 1992.The first pictures: perceptual foundations of Paleolithic art. *Perception.* 21:389-404.

Harries, M. H., Perrett, D. I., Lavender, A. 1991. Preferential inspection of views of 3-D model heads. *Perception.* 20: 669-680.

Hegdé J. and Felleman D. J. 2007. Reappraising the Functional Implications of the Primate Visual Anatomical Hierarchy. *The Neuroscientist.* 13(5):416-421.

Hochstein, S. and Ahissar, M. 2002. View from the Top: Hierarchies and Reverse Hierarchies in the Visual System. *Neuron.* (36): 791–804.

Hodgson, D. 2018. Tracing Upper Palaeolithic Iconography: The Strange Case of Animal Tracks. *Global Journal of Archaeology &. Anthropology.* 3(4). doi: 10.19080/GJAA.2018.03.555620

Hübner, R. and Volberg, G. 2005.The Integration of Object Levels and Their Content: A Theory of Global/Local Processing and Related Hemispheric Differences. *Journal of Experimental Psychology Human Perception & Performance.* 31(3):520-41.

Hudson, S. C. 1998. The Hunter's Eye: Visual Perception and Palaeolithic Art. *Archaeological Review from Cambridge.* 15 (1): 95-109.

Kersten, D. and Yuilley, A. 2003. Bayesian models of object perception. *Current Opinion in Neurobiology*. 13:1–9.

Kirby, K. N. and Kosslyn. S. M. 1992. Thinking Visually. In, G.W. Humphreys (Ed.), *Understanding Vision* (pp. 71-85). Cambridge, Mass: Blackwell.

Kosslyn. S. M. 1994. *The Resolution of the Imagery Debate*. Cambridge, MS: Bradford.
Kosslyn, S. A., Flynn, R.A., Amsterdam, J. B. and Wang, G. 1990. Components of High Level Vision: A cognitive neuroscience analysis and accounts of neurological syndromes. *Cognition.* 34: 203-277.
Logothetis, N. K. and Sheinberg, D. L. 1996. Visual Object Recognition. *Annual Review of Neuroscience.* 19: 577-621.
Marsolek, C. J. 1995. Abstract Visual Form Representation in the Left Cerebral Hemisphere. *Journal of Experimental Psychology: Human Perception and Performance.* 21 (2):375-386.
—. 1999. Dissociable neural, subsystems underlie abstract and specific object recognition. *Psychological Science.* 10: 111-118.
Marsolek C. J. and Bergund, E. D. 2008. Dissociable neural subsystems underlie visual working memory for abstract categories and specific exemplars. *Cognitive, Affective, and Behavioral Neuroscience.* 8 (1): 17-24.
Mazard, A., Laou, L., Joliot, M. and Mellet, E. 2005. Neural impact of the semantic content of visual mental images and visual percepts. *Cognitive Brain Research.* 24: 423– 435.
McCarthy R. A. and Warrington, E. K. 1990. *Cognitive Neuropsychology: A Clinical Introduction.* San Diego: Academic Press.
McMenamin, B. W., Marsolek, C. J., Morseth, B. K., Speer, M. K. F., Burton, P. C. and Burgund, E. D. 2016. Conflicting demands of abstract and specific visual object processing resolved by frontoparietal networks. *Cognitive, Affective and Behavioral Neuroscience. 16* (3): 502-515.
Meng, M., Cherian,T., Singal, G. and Sinha, P. 2012. Lateralization of face processing in the human brain. *Proceedings of the Royal Society B; Biological Sciences.* 279(1735): 2052–2061.
Mithen, S. J. 1988. Looking and Learning: Upper Palaeolithic Art and Information Gathering. *World Archaeology.* 19 (3):297-323.
—. 1990. *Thoughtful foragers - a study of prehistoric decision making.* Cambridge: Cambridge University Press.
—. 1991. Ecological Interpretations of Palaeolithic Art. *Proceedings of the Prehistoric Society.* 57 (1):103-114.
Navon, D. 1977. Forest before the tree: The precedence of global feature in visual perception. *Cognitive Psychology.* 9:353-383.
Neiworth, J. J., Gleichman, A. J., Olinick, A. S. and Lam, K. E. 2006. Global and Local Processing in Adult Humans (*Homo sapiens*), 5 Year Old Children (*Homo sapiens*), and Adult Cotton Top Tamarins

(*Saguinus Oedipus*). *Journal of Comparitive Psychology.* 120(4): 323–330.

Orban, G. A., Claeys, K., Nelissen, K., Smans, R., Sunaert, S., Todd, J. T., Wardak, C., Durand J-B. and Vanduffel, W. 2006. Mapping the parietal cortex of human and non-human primates. *Neuropsychologia.* 44 (13): 2647-2667.

Peissig, J. J. and Tarr, M. J. 2007. Visual Object Recognition: Do We Know More Now Than We Did 20 Years Ago? *Annual Review of Psychology.* 58:75–96.

Peressotti, F., Rumiati, R., Nicoletti, R. and Iob, R. 1991. New evidence for the perceptual precedence of global information. *Acta Psychologica.* 77: 35-46.

Perrett, D. I., Oram, M. W. and Ashbridge, E. 1998. Evidence accumulation in cell populations responsive to faces: an account of generalisations of recognition without mental transformations. *Cognition.* 67: 111-145.

Perrett, D. I., Oram, M. W., Hietanen, J. K., and Benson, P. J. 1994. Issues of Representation in Object Vision. In, M. J. Farah and G. Ratcliff. (Eds.) *The Neuropsychology of High-level Vision- Collected Tutorial Essays.* (pp. 33-61). Hillsdale, NJ: Lawrence Erlbaum.

Pizzagalli, D., Lehmann, D. and Brugger, P. 2001. Lateralized direct and indirect semantic priming effects in subjects with paranormal experiences and beliefs. *Psychopathology.* 34: 75–80.

Poplin, F. 1984. Sur Le Profil Dorso-Lombarde des Bisons Dans La Nature et Dans L'Art Paléolithique. In, H. Bandi., W. Huber., M. R. Sauter. and B. Sitter. (Eds.), *La contribution de la zoologie et de l'ethologie à l'interprétation de l'art des peuples chasseurs préhistoriqùes* (217-242). Fribourg, Switzerland: Editions Universitaires Fribourg.

Ramachandran, V. S. and Blakeslee, S. 1998. *Phantoms in the Brain - Human Nature and the Architecture of the Mind.* London: Fourth Estate.

Robertson, L.C., Egly, R., Lamb, M.R. and Kerth, L. 1993. Spatial attention and cuing to global and local levels of hierarchical structure. *Journal of Experimental Psychology: Human Perception and Performance.* 19:471–487.

Saito, A. Hayashi, M., Takeshita, H. and Matsuzawa, T. 2014. The Origin of Representational Drawing: A Comparison of Human Children and Chimpanzees. *Child Development.* 85(6): 2232–2246.

Sasaki. Y., Hadjikhani, N., Fischl, B., Liu,. A.K., Marrett, S., Dale, A.M. and Tootell, R.B.H. 2001. Local and global attention are mapped retinotopically in human occipital cortex. *Proceedings of the National Academy of Science USA*. 98:2077–2082.

Sereno, M.I., Dale, A. M., Reppas, J. B., Kwong, K. K., Belliveau, J. W., Brady, T. J., Rosen, B.R. and Tootell, R. B. 1995.Borders of multiple visual areas in humans revealed by functional magnetic resonance imaging. *Science.* 268 (5212):889-893.

Suzuki, K., Roseboom, W., Schwartzman, D. J. and Seth, A. K. 2017. A Deep-Dream Virtual Reality Platform for Studying Altered Perceptual Phenomenology. *Scientific Reports.* 7(1):15982.

Taylor, L. M. and Mitchell, P. 1997. Judgements of apparent shape contaminated by knowledge of reality: Viewing circles obliquely. *British Journal of Psychology.* 88: 653-670.

Thouless, R. H. 1931. Phenomenal Regression to the Real Object. *British Journal of Psychology.* 21:339-359.

Turnbull, O. H. 1999. Of Two Minds About Two Visual Systems. *Psyche* (on-line journal) 5 (8). http://journalpsyche.org/files/0xaa68.pdf

Turnbull, O.H., Carey, D.P. and McCarthy, R.A. 1997. The neuropsychology of object constancy. *Journal of the International Neuropsychology Society.* 3:288-298.

Warrington, E. K. 2009. Warrington and Taylor's 1978 paper: Author’s update. *Perception.* 38:933-947.

Warrington, E. K. and James, M. 1988. Visual apperceptive agnosia: a clinico-anatomical study of three cases. *Cortex*: 24: 13-32.

Warrington, E. K. and Taylor. A. M. 1978. Two categorical stages of object recognition. *Perception.* 7: 695-705.

Weissman D.H. and Woldorff, M.G. 2005. Hemispheric Asymmetries for Different Components of Global/Local Attention Occur in Distinct Temporo-parietal Loci. *Cerebral Cortex.* 15(6):870–876.

# CHAPTER FIVE

# SEEING THE UNSEEN

> When we view an image, we can usually understand what is displayed very quickly and easily. One particularly compelling example of this ability comes from animal categorization tasks in which participants respond whether or not there is an animal in the scene. (Zhu et al. 2016, p. 1)

## Brain modules for detecting animals and natural things

Recent brain imaging techniques have shown that there is a discrete area of the human temporal cortex devoted to natural things. Within that area there are further regions dedicated to encoding animals (Caramazza and Shelton, 1998; Caramazza and Mahon, 2003) (See Figure 5-1 and 5-2). Given the preceding remarks, and on the balance of evidence, the fact such an area exists makes sense. As a higher-order module in the temporal lobes, this suggests that long standing evolutionary determinants, as defined, will have been a major contributory factor. This is reflected in similar areas identified for encoding faces and tools (Caramazza and Mahon, 1998). Those areas did not appear by accident, as they were highly relevant to the survival of our distant ancestors, which is supported by the fact that, although modern humans have minimal contact with animals, the same area of the brain prevails. Moreover, similar areas for cars or motorbikes are absent, because they were not adaptively important during human evolution. Accordingly, animals, though largely irrelevant to the lives of most modern humans, still have a sizeable presence in today's brains. This informs us such areas are difficult to eradicate because they are deeply rooted in the cortex and would take many generations of "reverse" evolution to erase.

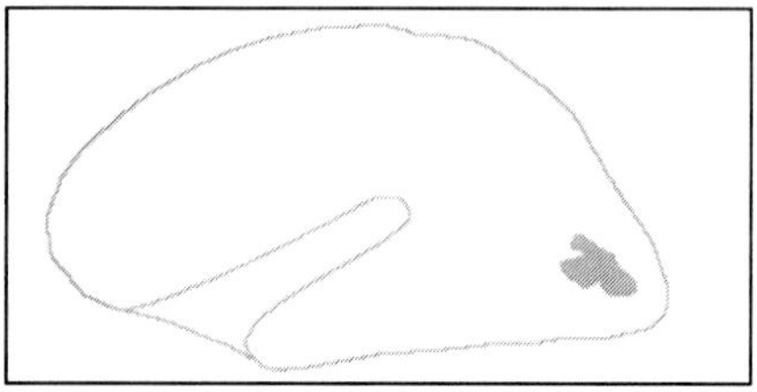

Figure 5-1. Area of human brain specializing, and most active, in processing animals (Adapted from Figure 3a in Mahon et al. 2009). Inflated brain map with frontal cortex to the left.

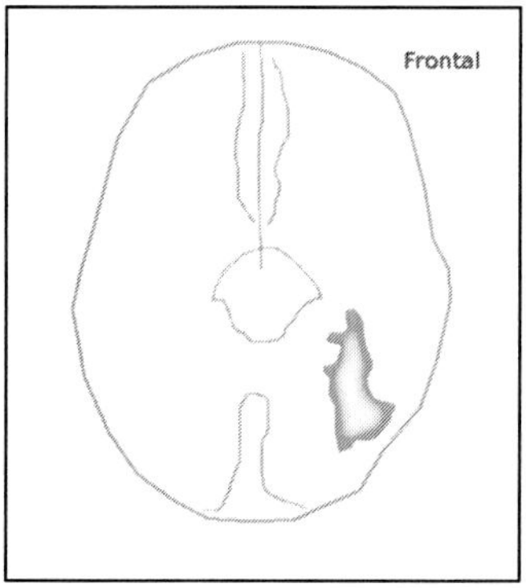


Figure 5-2. Underside of brain showing area specialised, and most active, for identifying animals. (Adapted from Figure 3 in Mahon and Caramazza, 2009).

Phobias provide further evidence as to the enduring preoccupation with animals as it is no coincidence that one of the commonest phobias concerns animals. In such cases, little inducement is required to provoke a phobic response, even though there may have been no previous aversive experience with animals (Seligman, 1972). Such reactions are easily induced in normal individuals in that only one paring of an animal with a painful event, such as an electric shock, can produce a phobic response; a reaction difficult to achieve with objects such as houses or flowers (Öhman et al. 1975a, 1975b). Moreover, in contrast to other stimuli, a degraded stimulus representing an animal triggers a phobic reaction via subcortical, emotional pathways (Öhman, 1993). Öhman (1986) makes the point that we fear animals because predation pressure was a primary force shaping our gene pool throughout mammalian evolution. As reactions must be rapid and automatic, with little inhibitory influence, the response tends to have an unconscious component to minimal input that persists (Öhman, 1975a). In addition, LeDoux (1994) found that such emotional driven memories are difficult, if not impossible, to erase, which is an important point because it is reflected in the long term preoccupation of humans with animals in art that seems equally difficult to "erase".

Those findings dovetail with our previous observations that the emotional brain can be tuned to particular things without conscious awareness intervening. Individuals suffering blindsight provide added confirmation. Although unable to explicitly identify objects because of deficiencies in the visual cortex, namely V1 (primary visual cortex), those suffering from blindsight can, nevertheless, when asked to guess, give the correct response to certain kinds of moving stimuli (Weiskrantz, 1996; Gaglianese et al. 2012). The detection of movement, however slight, is an important indication of the presence of an animate creature. Indeed, many animals remain highly sensitive to motion but experience difficulties in identifying stationary objects. Humans, however, appear to have the best of both worlds in being sensitive to both movement and static objects, the latter which facilitates the detection of stationary animals in camouflaged or degraded viewing conditions (Vyshedskiy, 2014).

A further example of animals easily springing to mind comes from hallucinations. Research has found the commonest objects experienced by those prone to visual hallucinations consist of animals and humans (Siegel, 1980; Shanon, 2003; Slade and Bentall, 1998). Related to hallucinations is Bonnett's syndrome (Schultz and Melzack, 1991), caused by a deterioration of the retina—usually in old-age—leading to a loss of sight and an accentuation of activity in the intact higher visual system. Such individuals report vivid hallucinations, many which feature animals seen as if in the real world, usually of a cartoon-like quality.

Furthermore, in the Rorschach Inkblot test (discussed in detail in a later chapter) one of the main categories reported is animals. The effect may be due to the projection of preferences onto a shapeless configuration often interpreted as a particular animal. Related to this, one study (Eagle et al. 1966) found that, when individuals were unaware of an animal embedded in a larger figure, when later prompted, the individuals concerned showed that something about the animal had, in fact, been registered. That effect was achieved by the clever ploy of arranging the branches of a tree to "accidentally" resemble the outline of an animal. Priming of this type has subsequently been confirmed by other research (Peterson et al. 2012), and similar studies of how priming surreptitiously influences behaviour indicate that the phenomenon is quite pervasive (Kouide and Dehaene, 2007).

By observing human behaviour in this way, we can be relatively certain that, along with the evidence already presented, this lends credence to the notion that our brains have an evolutionary-defined propensity, not only for the recognition of animal forms, but also for the projection of such forms onto objects that might fortuitously resemble animals. In this regard, Gombrich's (1960) suggestion the animals portrayed on the cave walls of France and Spain derive from a propensity of the human mind to project images onto suggestive surfaces may not be too far from the truth. In fact, Gombrich's theory has been misinterpreted by most critics. What he actually stated in Art and Illusion is that UP people projected their hopes and fears onto any shape that remotely suggested animals that derived from a "state of tension" (Gombrich, 1960, p.91), as elaborated in the following passage:

> Perhaps the conditions of their lives encouraged the early hunters to look for animal shapes in sacred caves rather than to make animals, to scan the vague forms of patches and shadows for the revelation of a bison, much as the hunter must scan the dusky plains for the outline of the hoped-for-prey. (Gombrich, 1960. p.93).

The problem with Gombrich's theory, however, resides in the proposed progression from simple outlines to more complete/complex depictions of animals from the early to late UP. Because quite complex depictions of animals, as at Chauvet, occur almost from the outset, namely during the Aurignacian and Gravetian, such a cumulative progression is unsustainable. This suggests that once the ability to produce iconic images of animals by way of projected mental images had occurred, mastering the more complex naturalistic depictions developed fairly rapidly (for reasons to be explained, I refer to these projective images as "hyperimages"). An observation confirmed by the fact that at Chauvet the earliest depictions display a range of graphic formats from the relatively simple, which often incorporate natural features, to the highly complex replete with details.

The fact that animals dominate palaeoart suggests hyperimages—as a product of a brain predisposed to detect animals—were the determining factor, especially as the related phenomenon of pareidolia (see below) would have led to a wide range of objects being projected onto cave walls. Gombrich's proposition can, however, be accommodated within the present context, due to the fact that the evolutionary-defined brain module for detecting animals was reinforced by the daily experience of animals so that the tendency to project animal images was more likely in palaeo-people to the extent they preferentially projected animal outlines even onto fairly-uniform surfaces. Cave walls therefore provided just one particularly opportune scenario for the graphic realisation of animals, as fauna can be found depicted on portable objects and outdoors. Nonetheless, cave depictions provide a vital clue as to why palaeo-artist began depicting fauna.

## Insights from pareidolia

There is an universal tendency amongst mankind to conceive all beings like themselves, and to transfer to every object those qualities with which they are familiarly acquainted, and of which they are intimately conscious. We find human faces in the moon, armies in the clouds.... (David Hume, 1758, p.33)

In order to understand the relevance of projective hyperimages to the origins of figurative depiction we need to examine pareidolia. Pareidolia refers to the way the mind perceives a pattern where none exists, e.g. faces in clouds. The experience is part of a broader psychological phenomenon known as apophenia, which includes other senses such as sound and touch, defined as the propensity to see meaningful patterns or connections in random data (Conrad, 1958). Fortunately, brain imaging has begun to reveal the subtle processes involved in pareidolia. For example, random visual displays were found to activate the higher top-down prefrontal cortex (especially the inferior and medial frontal gyrus together with right anterior insula) as well lower-order inputs ascending from occipital areas (Liu et al. 2014). What is intriguing about Liu et al's (2014) study is that impoverished stimuli with little to no suggestive features were enough for observers to think they saw faces in the presented array. As the authors' state:

> Our findings suggest that human face processing has a strong top-down component whereby sensory input with even the slightest suggestion of a face can result in a face interpretation. This tendency to detect faces in ambiguous visual information is perhaps highly adaptive given the supreme importance of faces in our social life and the high cost resulting from failure to detect a true face. (Liu et al. 2014, p. 76)

Another interesting finding of the study is that when subjects were primed to expect to see faces in random noise patterns devoid of any suggestive cues, similar to Figure 5-3, respondents continued to see faces. An observation that confirms the potency of top-down projective imagery to impose likely possibilities on irregular patterns when expectations are aroused, which is precisely what leads to projective hyperimages rather than mere pareidolia.

Figure 5-3. Random Array. (Modified from random noise image. GNU Free Documentation License, Creative Commons Attribution-ShareAlike 3.0 License. Author: IndiePhunq) https://en.wikipedia.org/wiki/File:RandomBitmap.png

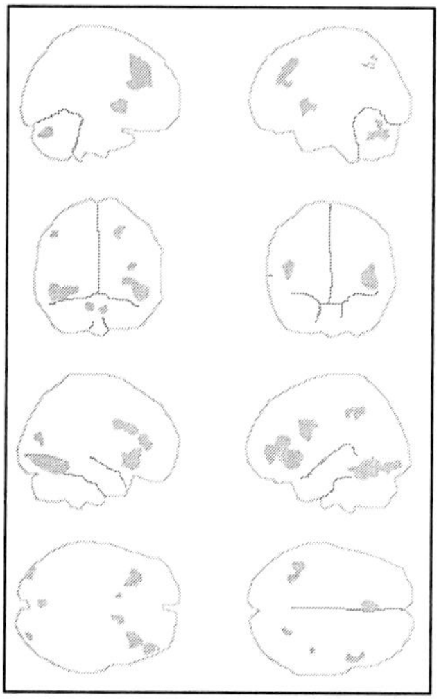

Figure 5-4. Brain areas activated in "pareidolia" (after Liu et al. 2014). From top to bottom, left to right: Medial (right hem. and left hem.). Rear View and Frontal View. Right Sideways View and Left Sideway View. Underside View and Top View.

The main region modulating face "pareidolia" was found to be the right fusiform gyrus, an area specialising in faces—both real and imaginary (Figure 5-4)—that serves as an interface between the inner and outer worlds. Liu et al also found specific activity in the primary visual cortex in response to images containing only contours or colours, yet frontal and occipitotemporal areas became active when subjects believed they could discern a face, namely areas that deal with high level imagery (see figure 5-4).

Those results suggest the fusiform face area is decisive for processing both illusory and real faces in that it serves as an interface where bottom-up information and top-down signals from the prefrontal cortex interact. In the present context, such observations can be applied to understanding the depiction of animals in palaeoart. If we replace "faces" with "animals" and "social" with "everyday" in the above quotation, the relevance becomes obvious. Thus,

> This tendency to detect animals in ambiguous visual information is perhaps highly adaptive given the supreme importance of animals in our everyday life and the high cost resulting from failure to detect a true animal.

Similarly, if we replace the fusiform face area with the part of the cortex for identifying animals (also partly in the lateral fusiform area), as illustrated in Figures 5-1 and 5-2, a similar scenario unfolds. The cortical areas encoding animals are thought to be innately defined for reasons already outlined as well as ongoing evidence from neuroscience (see Mahon et al. 2009).

It needs to be reemphasised that projective hyperimages are not the same as pareidolia, due to the fact that, unlike the latter, the former is not a random process as it relies on the primed hyper-charged visual system that conditions or biases visual experience (priming can also occur from cultural input). This is borne out by research demonstrating that those with religious or paranormal inclinations are more likely to see fictitious faces in visual arrays consisting of only the slightest hint of a face or even where no actual hint exists (Riekki et al. 2013). In other words, belief in the supernatural constitutes a form of self-priming, which is not dissimilar to the priming experienced by the UP hunter whose visual system was pre-tuned to see animals in amorphous patterns due to the need to detect fauna in order to survive. Ultimately, priming derives from the need to make sense of the visual world through predictions based on previous experiences by projecting suchlike onto the real world. This is especially useful when we are confronted with ambiguous situations involving subdued light or partially hidden forms.

Although the tendency for pareidolia is universal, because of its arbitrariness it is unable to account for the depiction of certain predators and prey in palaeoart, though it will have provided a platform for the more directed or targeted projective hyperimages. In the following section, we will show how pareidolia, projective hyperimages, a hunting lifestyle, and subdued lighting conspired to create the conditions for the production of the first graphic portrayals of animals. In order to achieve this, the significance of the Rorschach Inkblot and related perceptual phenomena need to be considered in greater detail.

## Clues from the Rorschach Inkblot

> It's better to mistake a boulder for a bear than a bear for a boulder. (Guthrie, 1993, p.80)

Ambiguous and challenging visual displays, such as the Rorschach Inkblot Rorschach (1942) and Dalmation Dog test are of particular interest because they provide a means of understanding how things as they appear in the real world relate to projective visual imagery. The Inkblot (Figure 5-5) consists of an amorphous configuration where, thanks to visual imagery, various objects are seen projected onto the display.

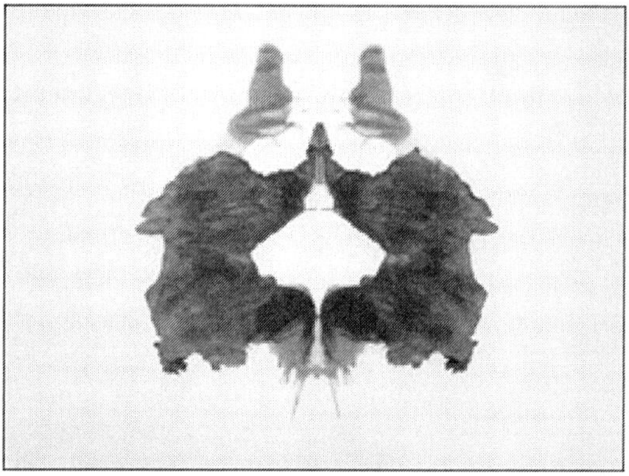

Figure 5-5. The Rorschach Inkblot.

Brain scans conducted by Asari et al. (2010) showed that when subjects see an image in a Rorschach Inkblot, the anterior prefrontal area, along with the amygdala (for emotion), become active (and artists tend to display more creative responses possibly linked to an enlarged amygdala). This suggests the matching of internally generated images with visual stimuli coming from the real world. Where these signals are fragmentary or ambiguous (as in the Rorschach Inkblot), incoming real world information passes through the occipitotemporal area to the temporopolar region where it is combined with top-down information (i.e., visual imagery) from the prefrontal cortex by way of the amygdala. This, in effect, "colours" the experience with emotion. The similarity with the Liu et al study is obvious and provides a crucial link between their findings on "pareidolia" and projective hyperimages in that the amygdala serves to drive the network already referred to so it is primed to look for and expect certain kinds of objects.

Research has discovered a striking similarity of fractals with Rorschach patterns (Taylor et al. 2017). Fractals are common in nature, and are particularly salient in the natural rock surfaces of caves and outdoor rock crags, and, when perceived, activate particular areas of the brain (Taylor and Spehar, 2016). It may be that fractal patterns automatically facilitate the search and find pattern detection system of the visual cortex for distinguishing forms in moderately noisy environments. What's more, fractals of moderate complexity are the most effective in stimulating pareidolia where animals and living things tend to predominate and, additionally, are the first entities to be perceived (Bies et al. 2016), as illustrated in Figure 5-6. In addition, when individuals were asked to draw their projective images, animal parts were frequently depicted.

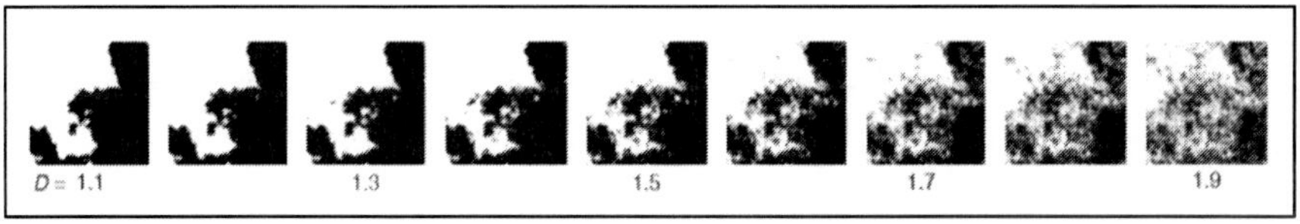


Figure 5-6. Fractal patterns 1.3 to 1.5 were found to facilitate pareidolia of living things more often and rapidly than simpler and increasingly more complex patterns (Thanks to A. J. Bies for permission to reproduce this image).

The extent of the existence of fractal patterns in caves containing UP animal depictions can be found in Figure 5-7.

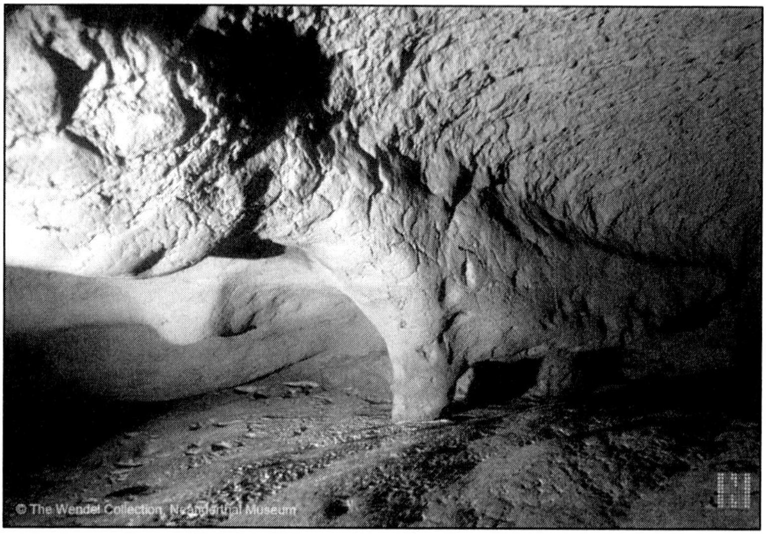


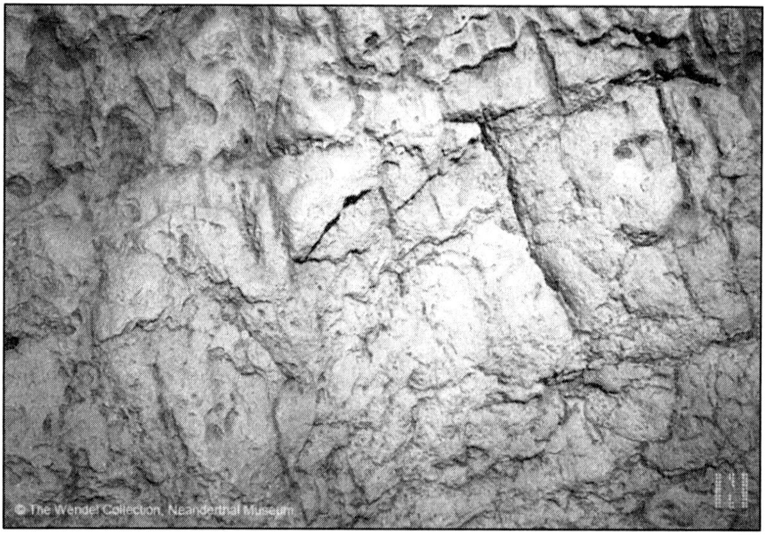


Figure 5-7. Ekain cave, France, which contains many animal depictions, showing a variety of different fractal patterns from two locations within the cave. The lower illustration also contains hints of depicted animals (©The Wendel Collection, Neanderthal Museum).

The relevance of the Rorschach Inkblot to UP art is reinforced by the fact it has been found to activate the mirror neuron system in a way that induces a feeling of movement that interacts with the corresponding system for touch (Girimoni et al. 2010). That interaction provides a link between the suggestive cave features (implied Rorschach) and active tactile exploration of rock surfaces through "palpation" as suggested by Pettitt et al. (2014). "Palpation" can be interpreted as a form of apophenia in the sense that a palaeo-artist may think that parts of animals are being touched as they move through the intense darkness of the caves. The depiction of animals in palaeoart may similarly be a way of bringing the cave walls to life by "touching" the surface with charcoal, ochre, and by way of engraving. In sum, caves and rock surfaces can be regarded as simulating the Rorschach Inkblot with many of the surfaces approximating fractals of 1.3 to 1.5, which are the most effective in evoking hyperimages of animals.

Similarly with the Dalmatian Dog test (Figure 5-8), where a challenging graphic array is employed to assess areas of the brain that become active when an observer suddenly differentiates a camouflaged dog from distractors (Ludmer et al. 2011). The amygdala was found to be particularly active together with the Lateral Occipital Complex (LOC), in the midlevel visual cortex, as well as the medial frontal cortex. Interestingly, LOC is where outline contours and symmetry undergo integration (Bona et al. 2014).

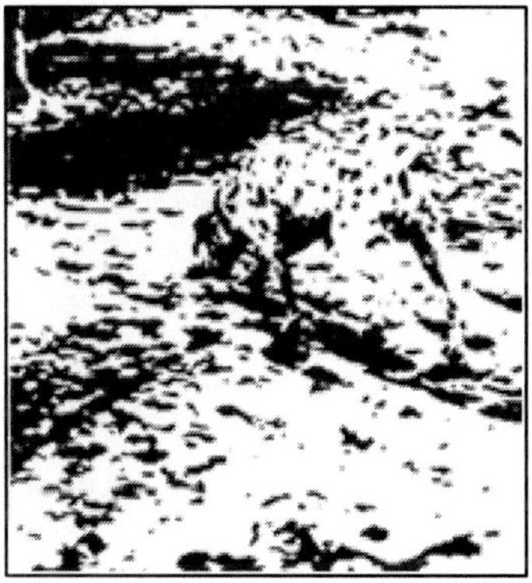

Figure 5-8. The Dalmatian Dog Image (original photo by R.C. James).

Ludmer et al concluded that the sudden, internally-mediated, solution to a perceptual problem induces long-term memory benefits because activation of the emotional areas helps consolidate information. In other words, a sense of reward/satisfaction is induced probably by the release of dopamine into neural channels (Petchkovsky, 2008). A sense of gratification is confirmed by the finding that when an individual disambiguates—through induced insight—an amorphous graphic array, equivalent to camouflage (similar to a Rorschach Ink Blot) in which an object is embedded, the brain's reward system becomes activate leading to an "Aha!" moment in successfully discerning an object (Ludmer et al. 2011). Fundamentally, once the Dalmatian has been identified it is difficult not to see the animal (the global precedence effect). An observation that suggests that when an ancient hunter-gatherer mistook a natural cave feature for an animal, the image summoned up in the "mind's eye" is likely to have persisted. A striking example of projective hyperimages, primed by ancestral determinants, can be found in a modern example shown in Figure 5-9. Note once the "animal" has been perceived it is difficult, if not impossible, to ignore the projected image due to the global precedence effect, which, you may recall, is particularly prominent in humans.

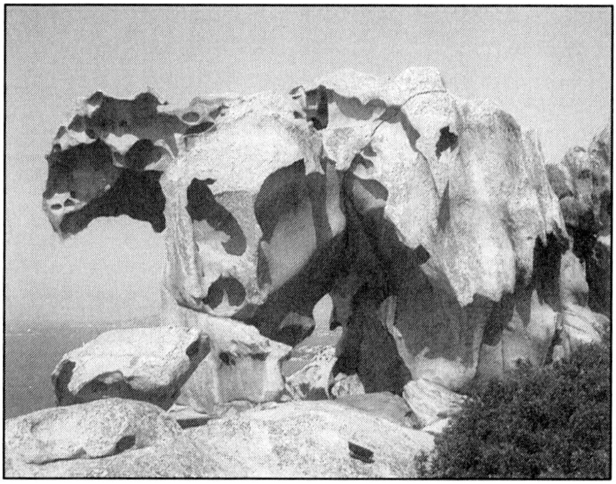

Figure 5-9. Capo d'Orso (Bear's Cape) near Palau (Sardinia). (Photo:Tobias Helfrich Creative Commons Attribution-Share Alike 2.5 Generic).

A similar, equally impressive, example can be found at Tito Bastillo Cave, Spain, where the evocative wall contours suggesting a bison have been accentuated by applying red paint (grey dots in Figure 5-10). A further example comes from Ekain Cave where the natural edge of a rock forms the back and tail of a bison (Figure 5-11).

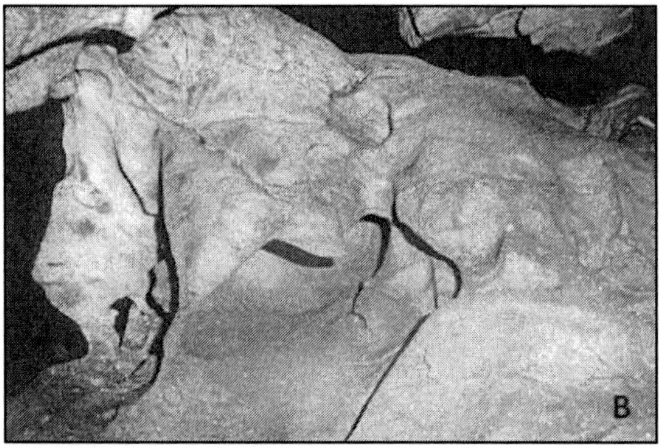


Figure 5-10. Tito Bustillo Cave showing a probable bison that takes full advantage of the natural wall relief. Note, the bison seems to have been marked out as an hyperimage by the insertion of the dark spots, which are red in the original image (Thanks to Rodrigo de Balbín-Behrmann for permission to reproduce this figure) [Original article: Figure 12B in, de Balbín-Behrmann, 2017].

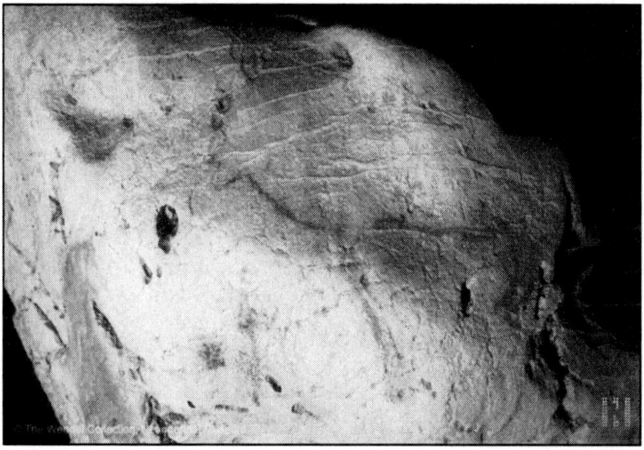

Figure 5-11. Ekain Cave showing a fine example of the exploitation of the edge of a protruding rock to create the cervico-dorsal line and tail of a bison (© The Wendel Collection, Neanderthal Museum).

Lorblanchet and Bahn (2017) claim that the beginning of iconic depictions can be explained by the pre-eminence of peck marks that initially derived from the skills required to make cupules, which were progressively joined together to make contour outlines. However, projective hyperimagery would have been necessary even in that task as a projective mental template is essential in order to "see" the animal in the first instance as well as to fill-in the gaps between peck marks. As indicated earlier, compared to other higher primates, the ability to "fill-in" missing sections of a figure, such as in the Kanizsa triangle and Street figure (see Figure 5-12), is exceptional in humans, which may derive from the necessity to complete the contours of a camouflaged or partly hidden animal, again as a function of the global precedence effect. We also saw earlier how even human infants are able to perform this in a drawing task with ease whereas chimps find the same task difficult to impossible.

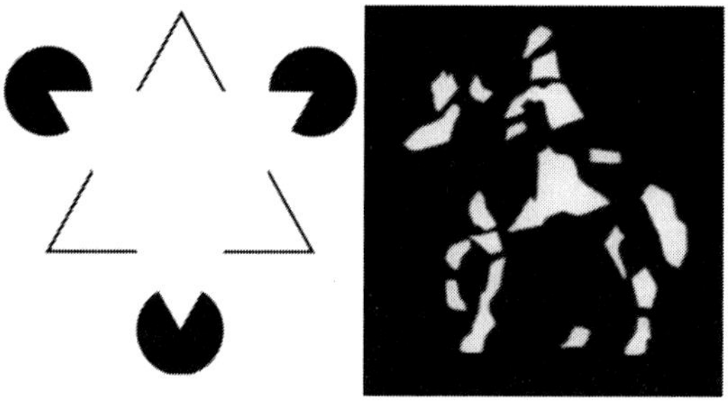

A B

Figure 5-12. A. The Kanizsa Triangle. Humans are able to see or "fill-in" the missing lines to complete the illusory triangle whereas chimpanzees require the existing graphics to be closer together. B. Street Figure. Example of how the human visual system is able to "fill-in" the missing contours to complete the horse and rider, which may derive from the need to disambiguate camouflaged or partly hidden animals.

Lorblanchet and Bahn's explanation also does not take into account the fact that animals were the main focus of concern over such a long period. Projective hyperimagery, however, is able to account for the preoccupation with animals as well as the formal ways they were represented. In fact, Lorblanchet and Bahn's approach is quite similar to Whitney Davis (Davis, 1986) who suggests that fortuitously arranged geometric marks or lines may have, on occasion, accidentally resembled an animal, which was then acted upon by accentuating the resemblance that could subsequently be copied and repeated. As it is possible to project a range of objects onto an array of peck marks or abstract lines, both approaches fall short of the mark. Nonetheless, projective hyperimages can provide an overarching framework by bringing the two theoretical positions together under the aegis of a single theory.

We saw that categorization is determined largely by the way the world is carved up according to the constraints of the visual system and one of the major ways this occurs is by classifying animals, which, at genus level, is the same cross-culturally (Atran, 1990; Berlin et al. 1973). Similarly, the way infants spontaneously begin to formulate categories provides the basis for the construction of later, more complex, classifications. Three-year-olds can, without prior learning, distinguish between the animate and inanimate according to certain telling features (Gelman, 1990; Broesch et al. 2014). Remarkably, research found that three to four-month-old infants are capable of discriminating cats, dogs and birds based solely on perceptual characteristics (Quinn et al. 1993; Mareschal and Quinn, 2001). That is probably why many children's' stories portray both benign and predatory animals that play on the infant's propensity to classify animals. Such input exploits a natural tendency in children that helps perfect the identification of animals at an early age. The fact infants are preconditioned in this way explains why animals are one of the most commonly perceived objects in the Rorschach Inkblot.

To summarise, the visual brain, as a result of the selective pressures experienced by successive generations of forebears, is superbly designed to detect different kinds of fauna, whether predator or prey. The ability is so acute that palaeo-humans were able to pick up cues about animals pre-consciously, a proclivity that begins "automatically" in infancy and undergoes refinement throughout childhood and beyond. That sensitivity is encoded in the visual system at various levels of elaboration, with the later stages of visual processing providing a template that can, through visual hypothesis testing, be matched to occasions where an animal might be particularly obscure. The short-cuts and tricks employed by the visual system therefore stood our ancestors in good stead when it came to deciding whether an animal was a gazelle, bison, or big-cat in various situations in the wild.

Although we have gone some way to answer the question as to how animals came to dominate palaeoart, this does not explain why the many other factors that figured prominently in the lives of Ice Age hunter-gatherers did not feature in depictions. The further question arises, how did the key-stimuli identified as crucial to the survival of our ancient ancestors come to be translated into representational depictions? In order to answer those issues we need to delve into how "representation" transpires in nature.

## Mimicry in nature and art

The use of "representation" in nature is everywhere. The representation of one thing for another is usually referred to as mimicry, which is not confined to the animal kingdom, as plants and insects employ the same device in a myriad of ways. There are several reasons why this trick evolved. It aids disguise, so a predator can remain unseen when approaching prey. Conversely, prey need to remain invisible to inquisitive predators. Both protagonists, therefore, mimic the surroundings they normally inhabit, usually by feigning the spots, dapples, and shadows of foliage, grass and trees e.g., black lines of tigers, the spots of leopards or the earthy hides of herbivores. Other obvious examples include the stick insect's uncanny resemblance to twigs, insects that look like leaves, and cuttlefish that mimic stones (see Figure 5-13). An interesting extension of this strategy is to be found in the Buff Tip moth, which disguises itself with markings on its wings, convincingly creating a 3D illusion of a broken twig—an example of three-dimensional perspective created on a two-dimensional surface (Figure 5-13). The moth, thereby, stands a better chance of survival. Many insects, like butterflies and moths, also have large eyespots on their wings making them look bigger and more fierce in order to deter potential attackers. Some flash the spots by rapidly opening and closing their wings, which brings the fake eyes to the immediate attention of any would-be assailant.

Mimicry may encourage a particular behaviour of mutual benefit to particular organisms. For example, orchids evolved to feign the markings, and sometimes the actual physical characteristics of insects, so that a male insect is duped into thinking the flower is a female insect and is attracted within to be dowsed with a generous helping of pollen. The insect, in return, receives the reward of nectar and is predisposed to return, while the orchid benefits through its pollen being widely distributed. Sometimes a plant or insect may resort to mimicry in order to gain from a creature's anti-predator defenses, such as toxins, without the cost of producing the actual toxin—referred to a Batesian mimicry. The benefits are twofold, avoidance of predators, and costly resources freed up and directed towards promoting survival.

Squid are masters of mimicry with the ability to change identity at will. Not only are these creatures capable of mimicking the immediate environment but can simulate fish with the ability to change form almost instantly. Reef squid are even capable of presenting two kinds of display at the same time. In fact, it has recently been established that squid, which have an unusually large brain, can learn to change bodily pattern in response to a reward. Professor Gilly of Stanford University achieved this in laboratory squid by conditioning them to mimic cross shapes demonstrating that some organisms are able to create displays independent of the natural environment. Other sea creatures, such as sponges and spider crabs, cover themselves in seaweed to blend in with the immediate surroundings. Sea dragons of Australia go one step further in actually looking like the seaweed in which most of their time is spent.

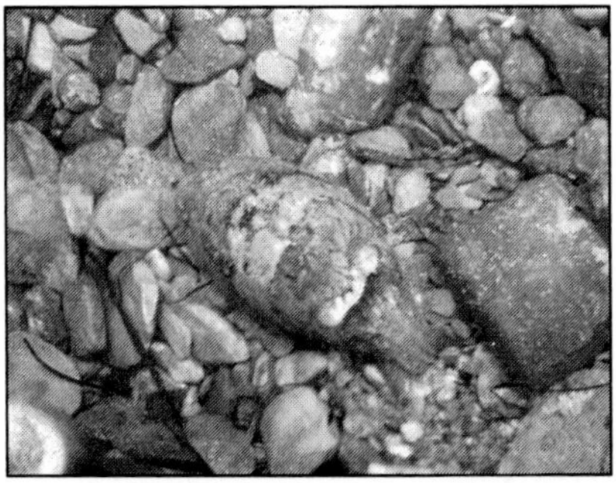

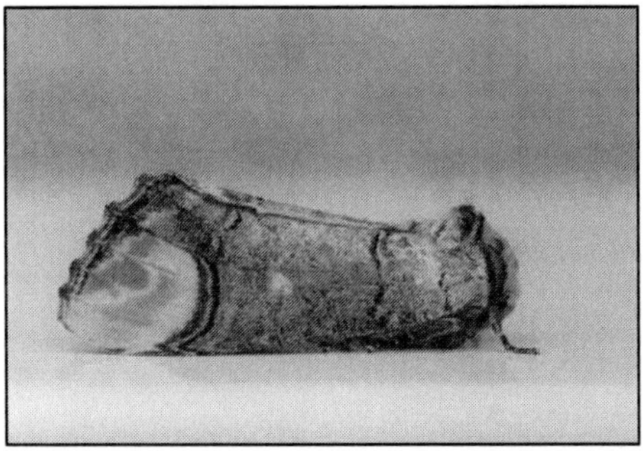

Figure 5-13. Examples of Camouflage and Mimicry. Top: Cuttlefish camouflaged amongst stones (Photo by Thomas Schoch. Creative Commons Attribution-Share Alike 3.0 Unported). Bottom: Buff Tip Moth mimicking a 3D twig (Photo by Böhringer Friedrich. Creative Commons Attribution-Share Alike 2.5 Generic).

What is intriguing about those examples is that humans often respond to mimicry in a similar way to the creature for which it was intended. Humans however, are able to see beyond the immediate resemblance, whereas animals are tied to particular trigger stimulus that releases a specific behavior. Yet, the very fact humans sometimes initially react to mimicry in a similar way to other species reinforces the notion that our visual capacities dovetail with the same determinants operating in the natural world. Nevertheless, humans benefit from a measure of flexibility and it is disentangling how this flexibility proceeds that is critical.

The fact mimicry can dupe us, though not always with the same behavioral outcome, says a lot about how human perception meshes with the same evolutionary determinants of other species. Importantly, this can work in reverse, as animals, such as chimps to dolphins, are able to perceive pictures produced by humans (Cabe, 1980;Bovet and Vauclair, 2000). Though whether untrained animals are able to see these as pictorial representations or mistake them—as human infants sometimes do (Winner, 1982)—for the real thing, remains to be established. In our case, we see the resemblance, and the difference, in a diverse range of instances and settings, whereas for most animals the response is usually linked to a particular stimulus trigger. Pertinently, research has found that different species react positively to objects portrayed in pictures according to whether the object is relevant to that species evolutionary niche (Cabe, 1980). Not surprisingly, when predators or potential mates are portrayed they are identified more straightforwardly than other objects.

It is arguable that squid have been manipulating mimicry to suit their own purposes for millions of years, which indicates that the ability to learn to use different kinds of "representational" cues in flexible ways is not entirely confined to humans. However, most animals, including the adaptable squid, are constrained by brain structures to do with emotionally responses issuing from the limbic system. Their signaling is, accordingly, closely aligned to their environment and is usually a product of sexual signaling or an attempt to remain concealed. The fact that humans are capable of interpreting mimicry more flexibly than animals suggests a capacity for manipulation unavailable to other creatures. The difference between squid and early humans in this context may reside in the large informational brain of the latter that allowed mimicry to be counteracted (Denys et al. 2005; Deacon, 1998). Mimicry in humans is, however, never completely overridden, as we are still prone to the same neurotransmitters as other primates in the face of certain images that suggest threat (LeDoux, 1994).

Mimicry in nature is a form of illusion because one creature is generating a resemblance to something to fool the sensory system of another into thinking that something continues to be the same when it is not. UP depictions can be regarded as a self-made visual illusion because they trigger the human visual system into thinking an animal is present where, in reality, none exists. This is because depictions trigger the same areas of the brain for the recognition of animals as occurs when animals are seen in the natural habitat. This is confirmed by the finding that the same region of the visual brain is activated on viewing an object in a photograph as when a line drawing of the same object is viewed (Grill-Spector et al. 2001; Kosslyn et al. 2001; Kreiman et al. 2000). Remarkably, when the same object is viewed in both a line drawing and a photograph, a similar level of activation occurs in the brain despite the reduced information in the former (Walther et al. 2011). Those findings suggest that line drawings provide valuable visual information about the world that the brain cherishes (Sayim and Cavanagh, 2011). Psychologists have exploited such preferences by producing clever illusions and bizarre effects. Most, if not all, of the illusions are thought to work through manipulating the evolutionary defined assumptions built into our visual systems at various levels of the visual cortex (Shepard, 1990).

Given the widespread utilisation of mimicry in nature, and our relative susceptibility to such pretence, it seems likely parts of the human visual system evolved to offset such effects. As deception and subterfuge were a pervasive strategy employed in the struggle to survive, humans may have coopted its heuristics for their own advantage. The learning and skills required for stalking and hunting necessarily involve deception, which may have been important to the evolution of abstract thinking in that previously learned information could be exploited to solve novel problems. As the animals that preyed on us and those on which we preyed were intimately involved in the battle of visual systems, it would be remarkable if our brains had not risen to this challenge. By incorporating the dynamics wrought from this battle of visual systems, palaeo-hunters benefited in becoming pre-adapted to perceive animals in challenging situations. Because such pre-adapted cues arose out of the cut and thrust of survival, it comes as no surprise that archaic humans came to create self-made cues that stimulated the same areas of the visual brain as when real animals were perceived. But instead, as in other creatures, of producing anatomically derived deceptive displays, the human response centred on expanding the brain to better co-ordinate the excessive information required to defeat such signals. As a result, rather than remaining passively susceptible to visual cues produced by other creatures, humans produced their own proactive cues. So, although squid can manipulate mimicry to fool predators and prey (and sometimes competing males), early humans sought to "fool" themselves by subverting and recruiting mimicry for their own ends thereby enabling the successful deception of animals on which human survival depended. In short, humans invented the ability to deceive themselves through graphic devices that mimicked the mimicry found in nature. Hence, unlike "lower" animals, humans came to devise strategies and tactics of stealth, such as animal disguises (of which more later), beyond those of the natural world thus making humans the supreme predator (Levy, 1999).

Deacon's (1998) idea that the procuring of meat is central to understanding the origins of human behavior is compatible with this idea. A notion supported by DeCasien et al's (2017) research mentioned earlier demonstrating the importance of diet to early hominin evolution. Deacon proposes that meat eating, as it became an important source of protein for *Homo habilis* onwards, led to crucial changes in the social fabric of foraging communities, involving a change in psycho-sexual bonding contingencies to do with the reciprocal negotiation between males and females, which centered on the procurement and sharing of difficult to procure protein. As well as being crucial to brain expansion, this change of diet promoted the use of tools that occasioned increasingly efficient ways of killing and butchering quarry (Shipman, 2010). So, in order to stay one step ahead of predator or prey,—instead of, as in other animals, becoming larger or more fierce-some (e.g. by evolving larger canines, claws etc.)—hominins "chose" the option of a larger, more complex, brain with all that this entailed for improved stalking skills. Paradoxically, just as early *Homo* lost the large primate canines, they invented stone tools that out-performed the lost incisors. As part of that loop, and in order to approach quarry using deception without being detected, the ability to mimic or "re-represent" the appearance of animals led to a further improvement in hunting prowess. In this connection, the use of disguise as a means of deceiving prey may have played a vital role in how such "representation" was to evolve, a ruse to which we will later return.

The discovery that represented features can trigger a behavioral response similar to when a real object is encountered, may have been critical to human cognitive/cultural development that marked a fundamental shift from a largely reflexive engagement with the world to a more proactive one. As Bednarik (2003a, p.127) states the capacity for iconicity, as manifest in the first representational depictions, is a "... 'managed' intentional use of visual ambiguity." Bednarik (2003b) further postulates, that mistaking a tree root for a snake on a forest path (recall the snake on the path and the fast alerting mechanism alluded to earlier) is an example of visual ambiguity that perhaps led early humans to see that one thing could stand for another. Once that notion had been understood, it was only a matter of time and successive approximation before fully-fledged 2D depictions were eventually created.

The realization that nature's capacity for mimicry could be subverted ultimately led—along with other crucial precursors to be discussed—to the first 2D Palaeolithic depictions. The preoccupation with depicting animals can, therefore, be ultimately traced back to the widespread exploitation of mimicry in nature and the way the visual brain of humans was fashioned by interaction with predators and prey (Hodgson, 2003a, 2003b).

Chimps, as higher primates, possess a limited ability for self-objectification, which was demonstrated by placing a spot on a chimp's forehead, after which it would view its reflection in a mirror. Chimps are able to realise that this is their reflection because they point at or attempt to erase the offending mark (Gallup, 1982). Most other primates are unable to achieve this and treat their reflection as another primate, which may be related to the fact that, in non-human primates, perspective-taking is limited due to a restricted theory of mind. That is, chimps are limited to not much more than level one intentionality (Schel et al. 2013) whereas humans can reach up to level five (Dunbar, 2003; 2007). Some primatologists challenge this finding but the fact remains that there are significant differences between humans and chimps. A deficit in levels of intentionality is reflected in the previous observation that, unlike human infants, chimps are unable to complete the missing lines of a drawn chimp face. Thus, humans are capable of, not just appreciating that a depiction represents an object, but also that other individuals are able to appreciate the same representation as such.

The ability to become aware of oneself as a separate entity provides a clue to how humans were able to exploit their self-image in ways unavailable to most other animals. In other words, that self-image became increasingly detached from the immediacy of the self, partly due to the social pressure of living in larger groups. One of the spinoffs of this, was the capacity to make animal disguises that, in turn, facilitated more effective strategies for stalking animals, as we shall see. Deacon (1998) suggests that the enlarged prefrontal cortex of humans is useful for exploiting information about something one might have just done or seen and reusing this to inhibit the tendency to immediately follow up a particular correlation and, instead, shift attention and action to an alternative association. Opting for the contrary to what is immediately available demonstrates a shift from a "natural" habitat response to one mediated by top-down rather bottom-up contingencies; an ability that requires overcoming the emotional pull of previously acquired reinforces. The enlargement of the prefrontal cortex, and the associated changes to connection patterns to other parts of the brain during evolution, introduced strong biases into learning that gave human prefrontal circuits a greater say in many behavioural responses. This is encapsulated in the earlier observation that humans are capable of overriding immediate responses to an alarming stimulus by imposing top-down constraints on lower-order reactions. A capacity that seems to be connected to the fact that humans have more neural feedback tracts that modulate lower order response systems compared to other higher primates (Eagleman, 2015). In essence, the ability to inhibit the normal response to naturally occurring deceptive signals and re-represent such cues for a range of purposes constituted an important threshold in the evolution of *Homo*; a tendency that began with a profusely interconnected brain (Changeux, 2017) that facilitated the capacity for "metarepresentation" (Leslie 1987, 1994), which eventually became externalised in depictive form.

In short, although most animals generally tend to mistakenly treat a representation as real, early humans eventually came to realize they could, to some extent, inhibit the associated automatic response as well as manipulate its heuristics to suit their own purposes. In other words, archaic humans began to exploit the mimicry employed by fauna to defeat the human visual system in the sense that it could be deployed for deceiving the same fauna through, for example, disguise that was later exapted for producing 2D depictions. What separated humans from animals, then, was the ability to imitate and dupe fauna by employing a more sophisticated assortment of mimicking skills, of which depictive representation may have been one further, but relatively late, development. *Homo* thereby turned the tables on nature by exploiting the trick of mimicry rather than, as in other animals, remaining subject to its dictates.

## The disguised hunter

The ability to mimic animals through disguise probably first emerged when humans began to populate the colder regions of the world with the need to keep warm by wearing animal skins. Hunter-gatherers up until quite recently employed animal skins, even in warm climes, to stalk animals (Thackeray, 2005) as can be seen in the San and Timucua people (Figures 5-14 and 5-15). The San of southern Africa also portrayed animal disguises in their rock art that had obvious mythic and ritualistic connotations. It would not have escaped the notice of UP people that the wearing of animal skins simulated various fauna and, on hunting expeditions, aided the safe approach to prey. A possible illustration of this tendency can be seen in Figure 5-16 and 5-17. Moreover, animal skins only require tweaking to appear yet more like the animals hunted, leading to the procurement of greater amounts of protein-rich meat with all the advantages this affords.

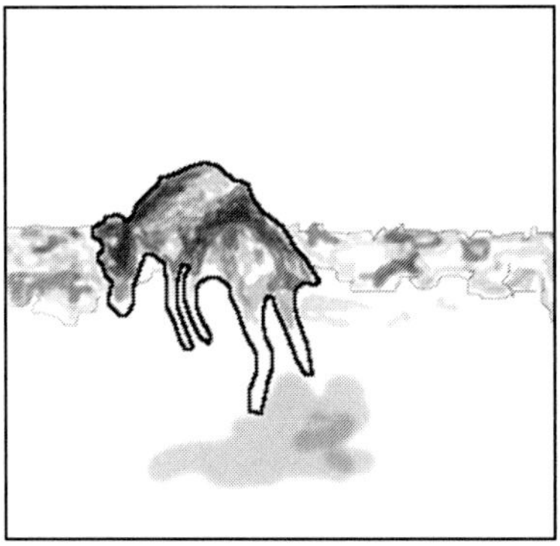

Figure 5-14. Illustration taken from a photograph by W.H.C. Taylor in 1934 of a human hunter (probably San) under the hide of a large antelope in the Kalahari desert, southern Africa (adapted from the original photograph by the author).

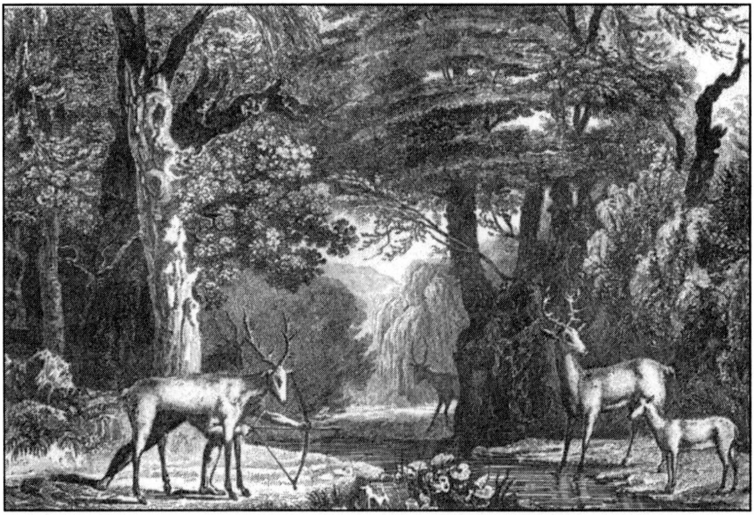

Figure 5-15. Drawing by Jacques Le Moyne of Timucua hunting deer in Florida, circa 1562. From the Florida Photographic Collection. (Creative Commons licence, Public Domain).

Pictorial representation was therefore most likely foreshadowed by other kinds of "representation" of which the wearing of animal skins, including upper features such as heads, antlers, horns and suchlike, will have helped hunters draw close to game without being detected thereby promoting a safer and more efficient mode of dispatch. Disguises employing predator-like features may have also been used to alert and stampede prey into traps and over cliff edges. Evidence for the use of animal disguises during the UP comes from the famous "lion man" sculpture that may be a person disguised as a lion ready to spook game into stampeding towards traps (Figure 5-16). Significantly, an UP cave in

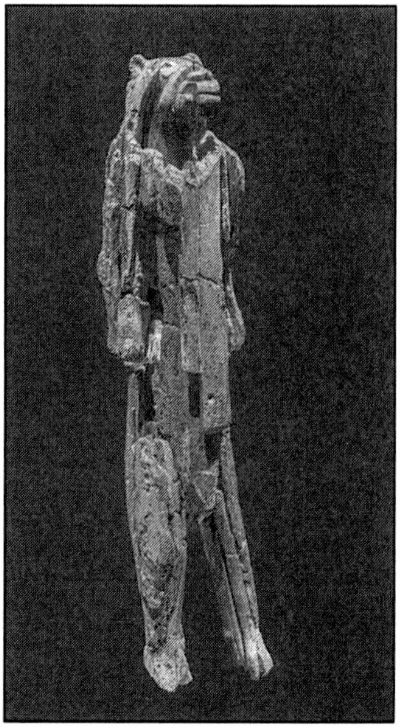

Figure 5-16. The Löwenmensch "Lion Man" apparently dating to at least 30,000 years ago (Image by Dagmar Hollmann / Wikimedia Commons. License: Creative Commons Attribution - ShareAlike 4.0 International).

Spain contains evidence of lion pelt exploitation dating to around 14,800 ago, which suggests such pelts may have been used for disguise (Cueto et al. 2016). The "lion man" figurine may also portray bear features, as bears often take on a standing posture and appear human-like from a distance. The painted figure (the "sorcerer") at Le Trois Frères cave may also represent a human dressed in animal skins wearing antlers that permitted hunters to approach game (Figure 5-17).

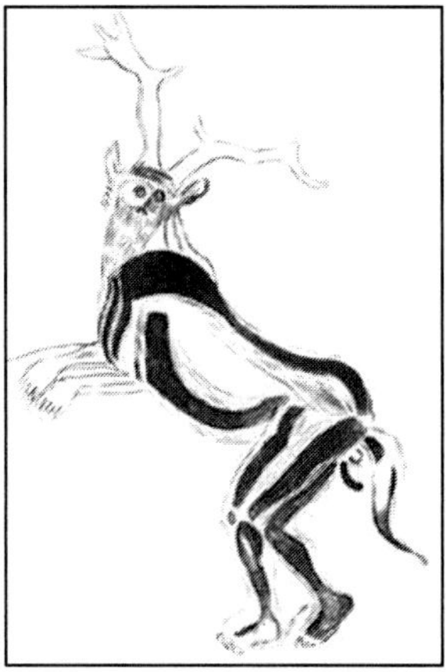

Figure 5-17. Drawing of possible disguised hunter based on an UP painting at Trois Freres Cave, France (drawing by H.Breuil).

In addition, a triangular cobble from La Madeleine (France) has engraved upon it a human figure in upright position with two heads, one thought to be a human, the other an animal-like mask (White, 2003). Further evidence comes from the semi-nomadic hunter/gatherer community of Star Carr, Yorkshire, England, dating to around 10,000 BP, where antler "frontlets" were probably worn during rituals (Laing and Laing, 1982) that could also have been used as a disguise for hunting animals (Conneller and Schadla-Hall, 2003) (see Figures 5-18 and 5-19).

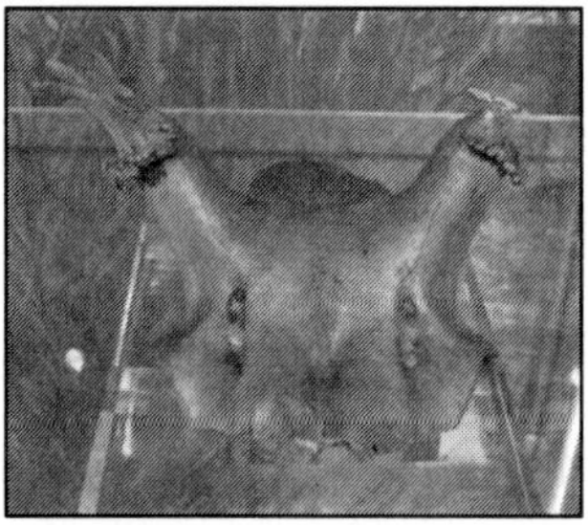

Figure 5-18. Antler frontlets from Star Carr (Photo by the author).

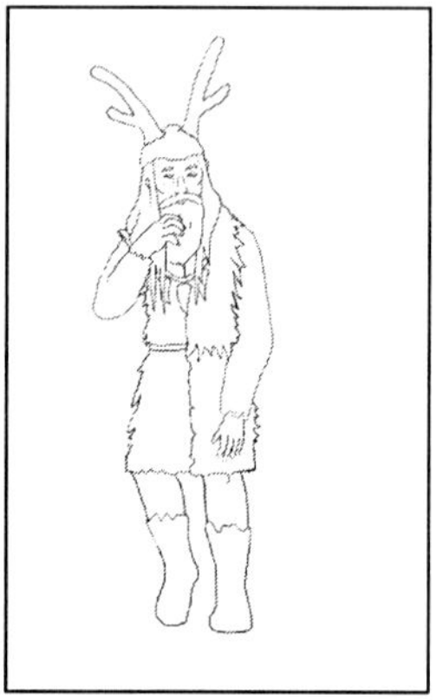

Figure 5-19. Artist's impression of how the antler frontlets illustrated in previous figure might have been worn (Adapted from original drawing by the author).

A therianthrope sculpture with the head of a bird and the legs of a human, thought to be around 33,000 years old, has additionally been found in southwestern Germany (Conard, 2003). Similarly, Manaseryan (2003) identified some human figures in rock art, dating to around 5,000 years ago, wearing masks in the context of scenes portraying animals being hunted. Though most of the above examples suggest ritual was important in such artworks, they were, nevertheless, probably foreshadowed by the ability to make and use animal disguises, which will have predated the two-dimensional representations of animals of the UP by a considerable period. In particular, because the ability to "re-represent" animals allowed a disguised hunter to approach fauna with relative impunity, but still accompanied by fear and awe related to a co-dependence with animals, this almost certainly led to the assigning of animals with supernatural power.

## Early "sculptures"

The long-term interdependence with animals, together with a larger informational brain, may have led to a propensity to attribute animal forms to natural occurring objects such as rocks, boulders, stones, etc., at a relatively early period. Sculptured objects may derive from such a proclivity in that the implicit iconicity of a rock or stone was accentuated by simply applying a few scratch or score marks. Such representations predate two-dimensional iconicity by several hundred thousand years, which suggests 2D depictions beginning around 40,000 years ago were not an isolated phenomenon but rather arose from a suite of previous but related representational activities, including animal disguises and mimicry, as the archaeological record is beginning to reveal.

In effect, there was probably a long period of passive appreciation of natural occurring objects that happened to resemble things such as faces, such as the Makapansgat cobble. This cobble is thought to have been collected by *Australopithecus africanus* and carried some distance to a cave habitation site. Mary Leakey also cites a stone from Olduvai Gorge, Tanzania, dating to around 1.8 million years ago that resembles a baboon's head (Leakey, 1971, plate 18). Those examples imply a "passive" or "reflexive" appreciation of natural objects that suggest faces or heads. A natural stone resembling a human body has also been unearthed at Berekhat Ram, Israel, dating to at least 233,000 years ago (Bahn and Vertut, 1997) inscribed with a few intentional marks delineating the main anatomical features. Similarly, the Tan Tan figure from Morocco, dating to around 400,000 BP (Bednarik, 2003b) portrays a human-like figure. A face-like rock, enhanced to accentuate the main features, has recently come to light in the Borzonasca region of Italy, which Pietro Gaietto (2014), who discovered the object, claims is over 200,000 years old.[1] Sheila Coulson also claims that a 70,000 year-old large rock outcrop in a cave—with signs of ritual activity—may have been seen as a snake by the occupants (Coulson et al. 2011). A rock artefact resembling a human or feline face, known as the Roche-Cotard "mask", possibly made by Neanderthals (recently confirmed as authentic [Lorblanchet and Bahn, 2017; Marquet and Lorblanchet, 2003]) and dating to about 32,000 BP and possibly 75,000 BP (Marquet et al. 2016), suggests archaic humans were capable of producing passive representations (see figure 5-20 and discussion on the artefact in Chapter 7). Such minimally enhanced objects I refer to as passive representations because they primarily rely on an accidental resemblance to real objects. This has led some (Marquet and Lorblanchet, 2003; Zilhão and d'Errico, 2000) to raise the possibility of a tradition of crude figurines existing well before 40,000 years ago.

"Passive" iconicity may be extremely ancient predating the appearance of modern humans but it is not as complex as "proactive" iconicity. Yet, it is logical that sculptures predated 2D graphics, as the creation of figurines from raw natural 3D objects is mentally easier to achieve than creating a 2D graphic equivalent (see Gombrich 2002). Two-dimensional representations require the translation of three-dimensions into two, which is more challenging compared to making sculpture. Having said this, it needs to be reemphasised that even the early three-dimensional figurines from Germany may have originally derived from projective hyperimages not least because many UP sculptural objects from France, such as those from Bédeilhac cave, derive from the provocative features of the natural stones/rock.

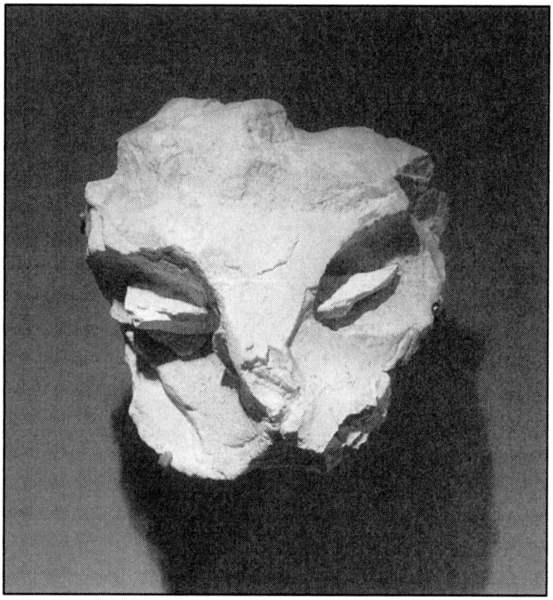

Figure 5-20. Neanderthal cat-like stone “mask” (photograph by author).

The sequence of events leading to representation, therefore, may begin with a "passive" or reflexive appreciation of found objects displaying a fortuitous resemble to animals or humans, which is in line with a guiding principle of psychology stating that comprehension usually precedes production (Nelson, 1977; Clark, 1993). Although those objects were at first collected and prized for their coincidental resemblance, much later in the archaeological record they were increasingly modified to become fully-fledged iconic objects. Nevertheless, not until around 40,000 years ago did representation become two-dimensional. An intermediary phase predating the UP may have, however, intervened where representation took the form of low relief representations but still based on natural evocative rock formations resembling animals and humans. This may be one reason many early examples of UP art incorporate the natural relief of cave walls. Once representation was laid bare in this way, its potential could then become the focus of cultural evolution.

As highlighted, the ability to depict animals will have coincidentally facilitated a sharpening of perceptual/recognition skills due to the intense concentration required, which is especially the case for cave depictions where visual thresholds were manipulated through exaggeration, occlusion, and distortion. A hypothesis corroborated by the finding that enhanced performance is experienced by artists in tasks involving perception and action understanding (Kozbelt and Seeley, 2007). By engaging in non-functional activities, early humans were able to gain a further edge in the struggle to outwit fauna leading to raised survival rates through a refining of the perceptual skills essential to the detection of animals.

Richard Coss (2017) makes a similar case but links visual memory to enhanced hand-eye coordination in humans compared to Neanderthals due to the latter's' inability to make and use refined tools for accurate throwing. According to Coss, humans benefited from advanced visuomotor control that interfaced with an exceptional visual memory that was essential to create figurative art. There are a number of problems with Coss's idea. First, contra Coss, refined wooden spears almost as streamlined as modern javelins were made by archaic humans 300,000 years ago (Thieme, 1997; Schoch et al. 2015), which may have been used both for hunting animals and warding off large sabre-toothed cats (Universität Tübingen Press Office, 2017). Despite their ballistic properties, it seems they were thrust at close quarters to animals rather than throne (Gaudzinski-Windheuser et al. 2018), which suggests that, similar to late Acheulean tools, they were over-refined for ongoing needs. Second, some recent hunter gatherers did not produce any figurative art. Rather, they relied on geometric motifs to express their artistic inclinations, which suggest representational competence cannot be employed as a marker of cognitive complexity. In addition, *Homo sapiens* did not produce fully-fledged representational art for 260,000 years; that is, since the arrival of the putative modern human brain 300,000 years ago (Hublin et al. 2017). Third, Neanderthals produced detailed geometric art 25,000 years before the UP figurate depictions (Hoffman et al. 2018), which suggests their visuomotor dexterity was not so limited. Lastly, recent research has found that Neanderthals hands are just as dexterous as modern humans (Karakostis et al. 2018). Having said this, visuomotor criteria may still be a factor to the extent that the cerebellum of Neanderthals appears to be smaller than that of both early and modern humans (Kochiyama et al, 2018), but this may relate to limitations in pro-sociality. Ultimately, the reason Neanderthals did not produce representational depictions may be linked to socio-cultural factors in that

*Homo sapiens* profited from growing population levels, mobility, the associated sharing of information, as well as cumulative material culture (Hodgson, 2013). It was not so much advances in visuomotor dexterity and visual memory that led to the depiction of animals, rather this was a coincidental outcome of an intense hunting lifestyle, an evocative environment and increased flexibility, which was coupled with an ability to share information leading to greater innovative outcomes.

It is commonly believed that the first representational art had no implications for survival. The story traced so far suggests otherwise. Because humans were now able to manipulate the key-visual stimuli of their own ecological niche, they gained the upper hand over the predators and prey populating that niche in the sense it was an "art" borne out of the need to see the unseen before the unseen detected humans. In this light, figurative art can be interpreted as a kind of technology that promoted survival (although it rapidly became much more significant for other reasons). A modern equivalent might be the ships and aircraft presented to seamen and airmen during the Second World War used for training recruits to recognise enemy craft—a procedure that required presenting a series of views, usually the side or front aspects of a ship, or the side and top of an aircraft.[2] Yet our ancestors seem to have stolen a march on those latter-day initiatives because palaeo-artists systematically degraded and depleted the representations of animals in order to improve perceptual thresholds. The apparent naturalism of UP depictions may well be a by-product of this procedure in that it derived from an evolved template of the visual brain specialised for the detection of animals. Even so, it should be noted that the depiction of animals and associated adaptive benefits were essentially accidental, as the visual brain evolved for the purpose of detecting animals, not for their portrayal. The fact that the same area of the visual brain is active when viewing a real animal as when a representation is observed, means that depictions exist in two worlds, that of the living and that of the fictive, yet the two modes are intimately connected. Precisely because of this two-fold nature, graphic images rapidly became the focus of other-worldly concerns.

## Pushing the recognition system to the limit

We already noted that palaeo-artists trimmed down the contours of animals in depictions to improve the ability to detect animals. You might ask the obvious question, how does this make the task of detecting animals easier when this should arguable have the opposite effect? Paradoxically, increasing the amount of information presented, leads to the visual system becoming more unresponsive. Conversely, by supplying less information, this gives the system more work to do, in a way that facilitates recognition. Less really does mean more, but the "less," in this case, is of a type that has to be portrayed in a particular way to work.

The phenomenon is effective because, as emphasized by neuroscientists (Kosslyn and Shin, 1994), imagery selectively facilitates perception, so that if subjects imagine an object they are more likely to detect a degraded version of the shape when presented with alternatives. In addition, the lower-level visual system can be "trained" by higher areas to recognise salient parts of a perceived object. By making the outlines of animals challenging, palaeo-artists were, albeit inadvertently, training the visual system to identify animals in degraded natural conditions e.g., camouflage, subdued lighting, etc. In this way, Palaeolithic depictions provided a surrogate for visual imagery by supplying a more reliable "sketch pad" for the detection of animals.

To give a practical example, when an insect or plant enthusiast searches for a particular species, a field-guide containing a detailed illustration of the creature is consulted. Even though this would normally be a fully fleshed-out representation, identification is, nevertheless, facilitated—otherwise the guide would not serve its function. The usefulness of manuals and guides may reside in the externalisation of visual memory that makes the finding of objects much easier. For instance, on discovery of the first example of a searched-for item, visual memory is stimulated, so subsequent items of the same type become easier to discern[3] because the initial find serves to prime, or grooms, the recognition system via visual memory. In effect, illustrations of the targeted objects serve as a substitute for the usually difficult-to-find first object, thereby saving on time and resources, which could—in more dangerous situations—prove costly. In sum, because visual memory is prone to fading due to storage limitations, illustrations serve as a convenient stand-in that bootstraps visual memory. The surprise comes in the discovery that by drastically reducing the amount of information in illustrations the task of finding even more difficult (perhaps obscured or camouflaged) items becomes less problematic.

That finding has been verified by perceptual psychologists, who established that depleted outline figures are more effective in tuning the visual system than more complete contour outlines (Snodgrass and Feenan, 1990). So much so, that moderately abbreviated outlines (just above recognition threshold) proved to be the most effective in identifying figures made of more depleted outlines that, in the normal course of events, would remain undetectable (see Figure 5-21). In short, identifying

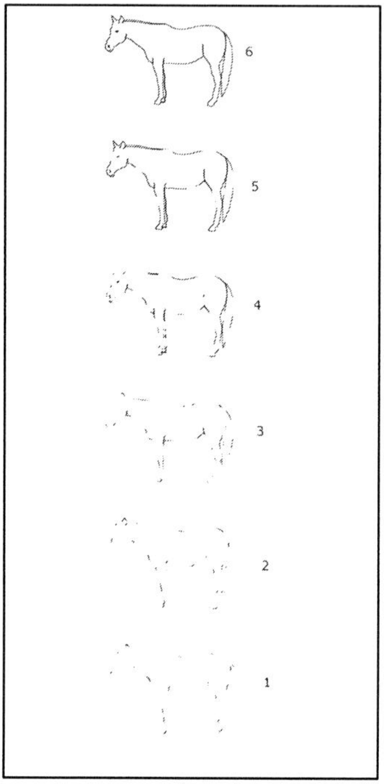


Figure 5-21. Animal outline contours showing the increasing depletion of contours across six levels of abbreviation. Level 3 is at recognition threshold and is the most effective for enacting perceptual closure including the recognition of more depleted levels. Note, stages above level 3 evoke a reduced amount of priming due to being more complete (drawing by author based on Snodgrass and Feenan's 1990 study).

animals at recognition threshold improves the detection of animals in challenging circumstances by facilitating perceptual closure, which has obvious benefits for hunters attempting to discern animals in camouflaged situations. In order to ensure visual memory is always geared up to its maximum, the most sensible strategy for an Ice Age hunter would have been to depict abbreviated outlines of animals in various situations. Although there are obvious examples of complete animals in palaeoart—in themselves more deplete than real animals—the abbreviated variety are more common precisely because they are much more effective than complete outlines in priming visual memory (Hodgson, 2003a).[4] This is supported by the fact that many UP portable objects and implements abound with illustrations of animals, which were probably transported on hunting expeditions to serve as a handy priming devise. As a result, most of the animal depictions in shelters, the open air and on portable objects, are depleted, and many are just above recognition threshold with a considerable number below this limit, as Figure 5-22, 5-23 and 5-24 (top) illustrate.

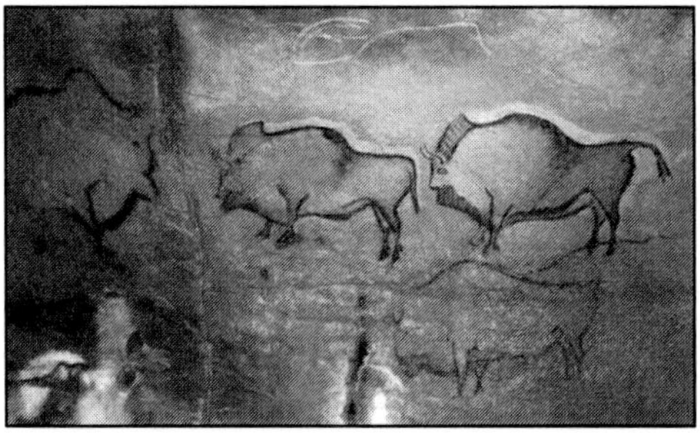

Figure 5-22. Four bison (perhaps five) depicted on the same panel at Covaciella Cave, Asturias, Spain with various degrees of depletion (Photo by José Manuel Benito, Public Domain).

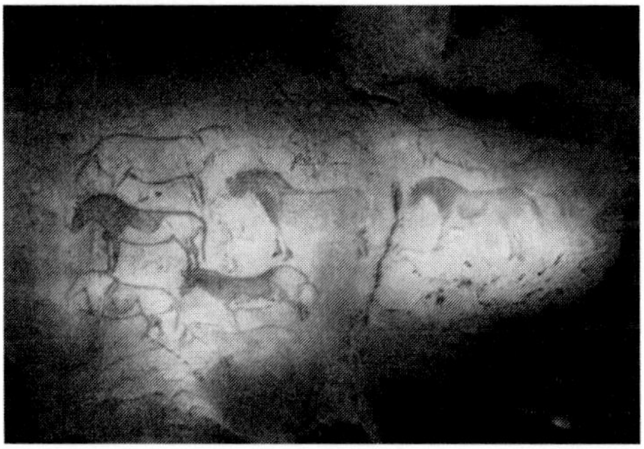

Figure 5-23. Ekain Cave, Spain, depicting horses and bison with various degrees of detail. Note some of the outlines resemble, and are difficult to differentiate, from the wall fissures and cracks. (Creative Commons Attribution-Share Alike 3.0 Unported).

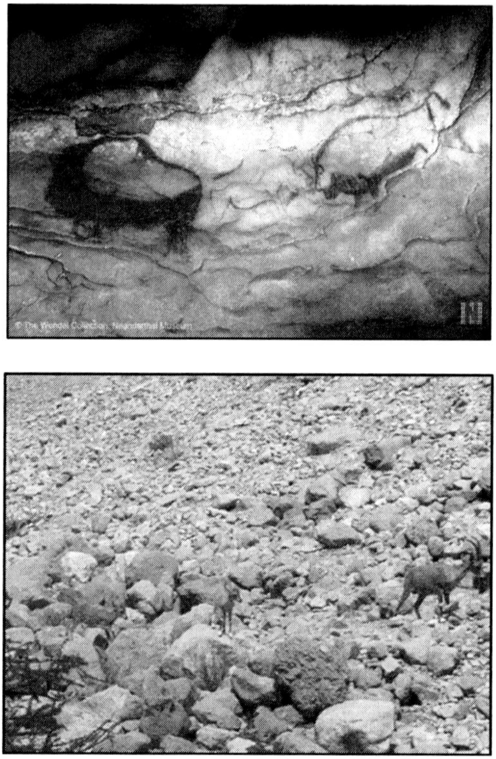

Figure 5-24. Top: Le Portel Cave. Three bison with various levels of depletion. The middle bison seems to emerge from the cracks in the wall (©The Wendel Collection: Neanderthal Museum). Bottom: In a similar way, the middle and last of the three ibex seem to emerge out of the natural rock environment (Creative Commons Attribution-Share Alike 3.0 Unported license).

By representing animals in abbreviated form, palaeo-artists increased their ability to detect animals on hunting forays because such animals would, more often than not, be encountered where outlines were difficult to discern. Figure 5-24 (top) shows how this scenario is simulated in an UP cave, where the natural cracks and fissures may have served as proxy camouflage in the sense that the animal outlines seem to progressively emerge from the suggestive cave features. Note how the middle bison is often overlooked by observers due its similarity with the natural fissures and eventually seems to emerge out of the wall cracks. The same procedure occurs in the lower figure in Figure 5-24 where the ibex seem to emerge progressively from the rocks in which they are camouflaged.

As the cave environment is one where viewing conditions varied, thanks to the utilisation of fires, torches and lamp burners etc., this had the effect of varying how well the depictions were discerned. Thus, the graphic techniques listed above were liable to further manipulation in ways that could either improve or degrade viewing conditions. The abbreviated depictions, together with the cave environment, hence provided an excellent theatre where the visual brain could deploy and explore recognition thresholds to the full.

Those pictorial devices may not always derive from a conscious pre-occupation. As noted, a wealth of pre-conscious processes may give rise to a conscious event. As animals were a significant force shaping the visual brain of humans throughout evolution, the likelihood of fauna becoming the focus of attention in the first depictive representations was inevitable—particularly as animals continued to figure prominently in the everyday life of UP people. Whether palaeo-artists deliberately resorted to producing depleted outlines to improve detection rates for animals we will never know for certain. Nevertheless, the fact they seem to have systematically manipulated such outlines is highly suggestive. The tests that perceptual psychologists employ to assess the capabilities and limits of the human visual system seem too similar to UP depictions to be a coincidence. That observation suggests palaeo-artists were deliberately using different graphic techniques as a rehearsal for the recognition of animal forms in different conditions. Notwithstanding this hypothesis, it seems certain that, even if they were not explicitly aware of the advantages accruing, the same outcome ensued i.e., improved recognition thresholds for detecting animals resulting in increased survival rates. A conclusion confirmed by the fact that trained artists are able to detect depleted, blurred and embedded objects better than none artists due to the advantageous gained from observational abilities and visuo-spatial/motor proficiency (Kozbelt, 2001).

The above findings demonstrate that UP depictions can be evaluated from a neuroscientific perspective. Nonetheless, such depictions almost certainly gave rise to supernatural associations stemming from the ability of representation to apparently bring animals to life in a way that simulated the emotions of fear and awe as much as when animals were confronted in real life.

The fact that visual imagery tends to improve visual memory (Pavio, 1971) suggests that palaeoart went one step further in providing benefits associated with the reliability that comes with a fixed image. Thus, as visual memory became externalised, a communally accessible interface was initiated with all the advantages this afforded. The perceptual principles governing the detection of animals were henceforth available for manipulation leading to a greater sense of control over the forces that shaped the lives of early humans, control that was, ultimately, to lead to the domestication of animals.

## Doubts and uncertainties

Some commentators reject the notion that the depictions of animals in palaeoart had anything to do with hunting or diet, as the bones found in caves, where meat was sometimes consumed, does not always match the animals portrayed, which led to the ultimate rejection of the hunting magic theory. Hunting magic theories propose that the UP depiction of animals were part of a ritual that brought good fortune on hunting expeditions. Given that there might be some justification for rejecting the hunting magic hypothesis, at the same time, this does not mean UP depictions had nothing to do with hunting on other counts. In fact, in caves such as Cosquer (France), there is a direct correspondence between the animals featured and the local fauna, including depictions of sea creatures, given that Cosquer was originally situated on or near the coast at the time the paintings were created (Clottes, 2000). Though it is true the dietary less important horses, bovines and mammoths are the most frequently depicted in caves, the more important sources of meat—such as reindeer—sometimes outnumber other animals on engraved limestone blocks (White, 2003), leading to the suggestion the animals selected for representation in caves need to be considered within a wider context.

As suggested, it makes sense that there was not always a direct correlation between the animals consumed and those depicted. Predatory animals will be predicted to feature in depictions, as well as species that do not fit easily into either category, because the visual brain's natural inclination to classify animals is a generalized one. Similarly, the inclusion of non-threatening animals in depictions is explained by the fact that in order to identify carnivores it is necessary also to identify benign fauna. What is more, wherever grazing animals are found carnivores are often lurking. For some animals, such as aurochs (wild cattle) and bison, category boundaries remain blurred, because, as well as being a source of food, as dangerous beasts, they also posed a threat. This is supported by the fact that Amazonian Indians, and those from technological societies with little or no contact with animals, when under the influence of psychoactive substances, commonly hallucinate all kinds of animals—

both fearsome and benign (Shanon, 2002). Nevertheless, there appears to be a significant amount of variation in the animals depicted over time. For example, Clottes (1996) found that in Chauvet there are a considerable number of predatory animals portrayed—as well as dangerous species such as rhinoceros. Notably, there are a substantial number of big-cats, compared to later caves, such as Lascaux, where horses prevail. Dowson and Porr (2001) make the further observation that a similar trend is evident in the statuettes of animals dating from more than 30,000 BP, where lions, bears and mammoths predominate. That tendency may reflect the fact carnivores and large animals, such as mammoths, suffered a drop in numbers during and towards the end of the UP. A similar drive towards the extinction of the same animals occurred in North America probably due to the colonization of that continent by Clovis people, though the changing climate may also have been a factor (Haynes, 2002).

There are deeper explanations why the correlation between what was represented and consumed is not perfect. Although most depictions are in caves, some in the most inaccessible reaches, this has led to a distortion of the archaeological record, not least because such locations are where conditions are most favorable to preservation. Actually, in recent times, a growing number of representations have been found outdoors on cliff-faces and even on the ground (Bahn and Vertut, 1997). Despite the fact many of the renditions are located deep in caves, palaeo-humans did not live in caves—though they sometimes temporarily occupied, ate, and made fires there. In any event, we can be reasonably sure that numerous portrayals of animals existed at open-air sites, which suggests the animals depicted in caves do not represent the total sum selected for depiction at any one time. Similar animal depictions, for example, can be found in great numbers on everyday objects such as weapons, tools, and plaques. Tellingly, the shafts of spear-throwers were often ornamented with particular fauna, suggesting animals were a common theme that co-existed in several contexts, of which we see only the surviving examples. The fact many of the portable objects on which animals are portrayed were made from the bones and antlers of fauna supports the fact there was an intimate relationship between the activities of UP hunter-gatherers and animals. The exploitation of animals for clothing, shelter, food, tools and weapons etc., together with the psycho/emotional criteria associated with hunting—involving stalking migrating herds—implies a complex interrelationship leading to different depictive outcomes according to the particular dynamic operating between the various factors outlined. In sum, though most scholars dismiss hunting-magic as an explanation, this does not negate the fact that hunting was important to the daily lives of UP people within the broader context of an interest in animals on a number of interrelated levels as indicated by Fritz and Tosello (2015).

Anthropologists and archaeologists have a tendency to look for and exaggerate the differences in palaeoart rather than the similarities in the search for identity markers and emblematic cues. In the quest to find ever more subtle nuances of cultural identity the slightest difference in graphic style is accentuated (Washburn, 2018). Such an approach emphasises the distinctive but comes with the danger of over interpretation, which is a perennial problem with interpreting prehistoric cultures.[5] As a result, many hypotheses, theories and speculations are put forward, which often play down or altogether ignore the obvious commonalities. Turning this debate around, an important question to be answered is why, during the 30,000 years of Upper Paleolithic art, did palaeoart not diversify into many radically different styles and themes instead of persisting with virtually the same way of depicting animals? Remarkably, despite large numbers of objects that can potentially be depicted in a variety of ways, palaeoart, throughout Europe including Russia, seems to have been almost exclusively concerned with the representation of animals and humans in predictable ways. There may be certain regional differences apparent but we should expect to find massive variations in subject matter, technique and modes of representation. This is quite remarkable, as it is obvious palaeo-artists had the necessary skills to represent a wide variety of objects in a host of different ways.

The recent finds from Creswell Crags, England (Bahn et al. 2003), as perhaps the most northerly outpost of European Upper Palaeolthic cave art, only serves to reinforce that view, as the depictions show strong similarities with cave art from the rest of Europe. Pertinently, many of the animal depictions from Creswell also derive from the predisposing natural rock features (Lorblanchet, 2007).

The preoccupation with animals, it should be noted, continued after the UP where, for example, Levantive Art shows a similar concern for hunting scenes (Criado-Boada and Penedo-Romero, 1993). Indeed, the tendency to portray animals in a similar way in rock-art can be found in Australian, Africa, North-America and countless other cultural milieus throughout "prehistory" as well as the historical period, and is prevalent in many myths and legends (Shepard, 1997; Lakoff, 1987; Ingold, 2000). The long-term concern for depicting animals in such a way has recently been confirmed by the discovery of a graphic representation of a "pig-deer" dating to around 35,000 years ago from Maros cave in Indonesia, (Aubert et al. 2014) as well as wild cattle from Lubang Jeriji Saléh cave

cave in Borneo dating to around 40,000 years ago (Aubert et al, 2018), which display similarities to the animals depicted in UP Europe.

## Notes

1.There is still controversy surrounding many of these objects with those supporting the "long-range" theory of derivation emphasising their authenticity, as opposed to those who take the "short range" or "sudden upsurge" view who regard them as flukes and therefore not representative of a general trend. In truth, some of the finds may be more genuine than others. See later in main text for more on this.

2. Interestingly, such training procedures gravitated towards using 3D models rather than simply relying on two-dimensional representations.

3. The expeditious eye is also important here because visual memory probably depends on such "icons" for storing and retrieving visual information. Accordingly, where this template decays over time there will be only a trace of the original image remaining in the underlying connectionist circuits. Beyond a certain threshold of decay, this trace will be incapable of being retrieved. However, in cases where the template has not completely decayed, seeing the “real” object in situ will be enough to bootstrap visual memory and therefore supply the missing neural links to generate the retrieved image. The procedure for immediate phenomenological experience of objects may be based on a similar scenario because commonly seen objects will, for short and longer term memory, become more firmly consigned to neural networks through familiarisation. Therefore, the time to bootstrap a "view" would be rapid and immediate because the full-blown icons relating to such objects have become well established in neural circuits.

4. Coincidentally, priming studies have shown that, whereas the typical view facilitates the recognition of nearby views of an object, unusual views facilitate similar unusual views but, in addition, automatically prime the typical view (Srinivas, 1993). Thus, the usual view springs to mind more frequently because it has widespread links to other views.

5. This raises an interesting philosophical question; at what point of refinement do we make a distinction that defines the appropriate level of analysis? The same criticisms that archaeologists make of those who take a more general approach to these issues can be levelled at archaeologists themselves, as it is always possible to examine a culture with an ever more detailed level of scrutiny. This problem is reflected in scientific research, where those espousing a broad theoretical position, which aspires to bring the threads of empirical research together under one banner, tended to be suspicious of those concerned with the particularities of the world. In contrast, those who analyse events at close quarters empirically tend to regard theoreticians with suspicion (Shapin, 1996). The two approaches, in fact, reflect

deductive and inductive forms of reasoning and are, ultimately, interdependent. What we have been endeavouring to demonstrate is how, from a generalised frame of reference, cultural diversity (of which artefacts are an indicator) can be approached with reference to that interdependence. However, archaeologists and anthropologists would raise the obvious defence that the theoretical level at which it would be valid to analyse a culture is defined by the actual culture itself. This amounts to little more than a tautology because it still implies that archaeologists and anthropologists have some "higher" frame of reference or authority that allows them to study, analyse and understand a culture in ways unavailable to the culture itself. It is clear that I take the view that archaeologists and anthropologists are engaged in the empirical study of human behaviour and artefacts from a wide range of communities that collapses into a much general theoretical context underscored by evolutionary theory, how the brain functions, and the psychology of perception.

## References

Asari, T., Konishi, S., Jimura, K., Chikazoe, J., Nakamura, N. and Miyashita, Y. 2010. Amygdalar modulation of frontotemporal connectivity during the inkblot test. *Psychiatry Research: Neuroimaging* 182: 103–110.

Atran, S. 1990. *The cognitive foundations of natural history.* New York: Cambridge University Press.

Aubert, M., A. Brumm, A., Ramli, M., Sutikna, T., Saptomo, E. W., Hakim, B., Morwood, M. J., van den Bergh, G. D., Kinsley, L. and Dosseto, A. 2014. Pleistocene cave art from Sulawesi, Indonesia. *Nature.* 514:223–227.

Aubert, M., Setiawan, P., Oktaviana, A. A., Brumm, A., Sulistyarto, P. H., Saptomo, E. W., Istiawan, B., Ma'rifat, T. A., Wahyuono, V. N., Atmoko, F. T., Zhao, J.-X., Huntley, J., Taçon, P. S. C., Howard, D. L. and Brand, H. E. A. 2018. Palaeolithic cave art in Borneo. *Nature.* 1476-4687. doi.10.1038/s41586-018-0679-9

Bahn, P. G., Pettitt, P. and Ripoll, S. 2003. Discovery of Palaeolithic cave art in Britain. *Antiquity.* 77 (296): 227-231.

Bahn P. G. and Vertut, J. 1997. *Journey through the Ice Age.* London: Weidenfeld and Nicolson.

Bednarik. R. G. 2003a. The Earliest Evidence of Palaeoart. *Rock Art Research.* 20: 3-28.

—. 2003b. A Figurine from the African Acheulian. *Current Anthropology.* 44 (3): 405-413.

Berlin, B., Breedlove, D. and Raven, P. 1973. General principles of classification and nomenclature in folk biology. *American Anthropologist.* 75:214-242.

Bies, A., Kikumoto, A., Boydston, C., Greeenfield, A., Chauvin, K., Taylor, R. and Sereno, M. 2016. Percepts from noise patterns: The role of fractal dimension in object pareidolia. *Journal of Vision.* 16(12): 790.

Bona, S., Herbert, A., Toneatto, C., Silvanto, J. and Cattaneo, Z. 2014.The causal role of the lateral occipital complex in visual mirror symmetry detection and grouping: An fMRI-guided TMS study. *Cortex.* 51:46-55.

Bovet, D. and Vauclair, J. 2000. Picture recognition in animals and humans. *Behavioural Brain Research.* 109 (2):143-165.

Broesch, J., Barrett, H. C. and Henrich, J. 2014. Adaptive Content Biases in Learning about Animals across the Life Course. *Human Nature* 25:181–199.

Cabe, P. A. 1980. Picture perception in nonhuman subjects. In, M. A. Hagen (Ed.), *The Perception of Pictures* (pp. 305-343). Vol.II. New York: Academic Press.

Caramazza, A. and Mahon, B. Z. 2003. The organization of conceptual knowledge: the evidence from category-specific semantic deficits. *Trends in Cognitive Sciences*. 7(8):354-361.

Caramazza, A. and Shelton, J. R. 1998. Domain-specific knowledge systems in the brain: the animate-inanimate distinction. *Journal of Cognitive Neuroscience.* 10: 1-34.

Changeux, J-P. 2017. Climbing Brain Levels of Organisation from Genes to Consciousness. *Trends in Cognitive Sciences*. 21(3): 168-181.

Clark, E.V. 1993. *The lexicon in acquisition.* Cambridge: Cambridge University Press.

Clottes, J. 1996. Thematic changes in Upper Palaeolithic art: a view from the Grotte Chauvet. *Antiquity*. 70:276-288.

—.2000. The "Three "Cs": Fresh avenues towards European Palaeolithic art. In, C. Chippindale, and P. S. C. Taçon (Eds.), *The Archeaology of Rock Art* (pp.112-129). Cambridge: Cambridge University Press.

Conard, N. J. 2003. Paleolithic ivory sculptures from southwestern Germany and the origins of figurative art. *Nature.* 426:830-832.

Conneller, C. and Schadla-Hall, T. 2003. Beyond Star Carr: The Vale of Pickering in the 10th Millenium BP. *Proceedings of the Prehistoric Society.* 69: 85-105.

Conrad, K. 1958. *Die beginnende Schizophrenie. Versuch einer Gestaltanalyse des Wahns* [The onset of schizophrenia: an attempt to form an analysis of delusion]. Stuttgart: Georg Thieme Verlag.

Coss. R. G. 2017. Drawings of Representational Images by Upper Paleolithic Humans and their Absence in Neanderthals Reflect

Historical Differences in Hunting Wary Game. *Evolutionary Studies in Imaginative Culture.* 16-38. doi: 10.26613/esic/1.2.46
Coulson, S., Staurset, S. and Walker, N. Ritualized Behaviour in the Middle Stone Age: Evidence from Rhino Cave, Tsodilo Hills, Botswana 2011. *PaleoAnthropology*. 18–61. doi:10.4207/PA.2011.ART42
Criado-Boada, F. and Penedo-Romero, R. 1993. Art, time and thought: a formal study comparing Palaeolithic and postglacial art. *World Archaeology*. 25 (2):187-203.
Cueto, M., Camarós, E., Castaños, P., Ontañón, R. and Arias, P. 2016. Under the Skin of a Lion: Unique Evidence of Upper Paleolithic Exploitation and Use of Cave Lion (*Panthera spelaea*) from the Lower Gallery of La Garma (Spain). *PLoSONE*. 11(10): e0163591. doi.org/10.1371/journal.pone.0163591
Davis, W. 1986. The Origins of Image Making. *Current Anthropology*. 27 (3): 193-215.
Deacon, T. W. 1998. *The Symbolic Species - The co-evolution of language and the human brain.* London: Penguin.
de Balbín-Behrmann, R., Alcolea-Gonzàlez, J-J. and Alcaraz-Castaño, M. 2017. The Palaeolithic art of Tito Bustillo cave (Asturias, Spain) in its archaeological context. *Quaternary International.* 430: 81-96.
DeCasien, A. R., Williams S. A. and Higham, J. P. 2017. Primate brain size is predicted by diet but not sociality. *Nature: Ecology and Evolution*. 1(011). doi: 10.1038/s41559-017-0112
Denys, K., Vanduffel, W., Fize, D., Nelissen, H.S., Georgieva., Vogels, R., Van Essen, D. and Orban, G.A. 2005. Visual Activation in Prefrontal Cortex is Stronger in Monkeys than in Humans. *Journal of Cognitive Neuroscience*. 16 (9): 1505-1516.
Dowson, T. A. and Porr, M. 2001. Special objects—special creatures: Shamanistic imagery and the Aurignacian art of south-west Germany. In, N. Price (Ed.), *The Archaeology of Shamanism* (pp. 165-177). London: Routledge.
Dunbar, R.I.M. 2003. The social brain: mind, language, and society in evolutionary perspective. *Annual Review of Anthropology*. 32: 163-181.
—. 2007. The social brain and the cultural explosion of the human revolution. In, P. Mellars, K. Boyle, O. Bar-Yosef. and C. Stringer (Eds.), *Rethinking the Human Revolution* (pp. 91-98). Cambridge: McDonald Institute Monographs.
Eagle, M., Wolitzki, D. L. and Klein, G. S. 1966. Imagery: Effect of a Concealed Figure in a Stimulus. *Science.* 151 (3712): 837-839.

Eagleman, D. 2015. *The brain: the story of you*. Edinburgh: Canongate Books.
Fritz, C. and Tosello, G. 2015. From Gesture to Myth: Artists' techniques on the walls of Chauvet Cave. In, R. White and R, Bourrillo (Eds.), with collaboration of F. Bon. *Aurignacian Genius: Art, Technology and Society of the First Modern Humans in Europe*. Proceedings of the International Symposium, April 08-10 2013, New York University. *P@lethnology*. 7: 280-314.
Gaglianese, A., Costagli, M., Bernardi, G., Ricciardi, E. and Pietrini, P. 2012. Evidence of a direct influence between the thalamus and hMT+ independent of V1 in the human brain as measured by fMRI. *NeuroImage* 60 (2)2: 1440-1447.
Gaietto, P. 2014. *Anthropomorphic Paleolithic Sculpture*. Morrisville, NC: LuLu Press.
Gallup, G. G. 1982. Self-awareness and the emergence of mind in primates. *American Journal of Primatology*. 2:237-248.
Gaudzinski-Windheuser, S., Noack, E. S., Pop, E., Herbst, C., Pfleging, J., Buchli, J., Jacob, A., Enzmann, F., Kindler, L., Iovita, R., Street, M. and Roebroeks, W. 2018. Evidence for close-range hunting by last interglacial Neanderthals. *Nature Ecology & Evolution*, 2018; 2(7): 1087-1092.
Gelman, R. 1990. First principles organize attention to and learning about relevant data: Number and the animate-inanimate distinction as examples. *Cognitive Science*. 14: 79-106.
Giromini, L., Porcelli, P., Viglione, D. J., Parolin. L. and Pineda. J. A. 2010. The feeling of movement: EEG evidence for mirroring activity during the observations of static, ambiguous stimuli in the Rorschach cards. *Biological Psychology*. 85(20): 233-241.
—. Gombrich, E. H. 1960. *Art and Illusion*. London: Phaidon.
—. 2002. *The Preference for the Primitive*. London: Phaidon.
Grill-Spector, K., Kourtzi, Z. and Kanwisher, N. 2001. The lateral occipital complex and its role in object recognition. *Vision Research*. (41):1409-1422.
Guthrie, S. E. 1993. *Faces in the Clouds: A New Theory of Religion*. Oxford: Oxford University Press.
Haynes, G. 2002. The Catastrophic Extinction of North American Mammoths and Mastodons. *World Archaeology*. 33 (3): 391-416.
Hodgson, D. 2003a. Seeing the "Unseen": Fragmented Cues and the Implicit in Palaeolithic Art. *Cambridge Archaeological Journal*. 13 (1): 97-106.

—. 2003b. The Biological Foundations of Upper Palaeolithic Art: Stimulus, Percept and Representational Imperatives. *Rock Art Research.* 20 (1):3-22.
—. 2013. Cognitive Evolution, Population, Transmission, and Material Culture. *Biological Theory.* 7: 237-246.
Hoffmann, D. l., Standish, C. D., García-Diez, M., Pettitt, P. B., Milton, J. A., Zilhão, J., Alcolea-González, J. J., Cantalejo-Duarte, P., Collado, H., de Balbín, R., Lorblanchet, M., Ramos-Muñoz, J., Weniger, G.-Ch. and Pike, A. W. G. 2018. U-Th dating of carbonate crusts reveals Neandertal origin of Iberian cave art. *Science.* 359 (6378): 912-915.
Hublin, J-J., Ben-Ncer, A., Bailey, S. E., Freidline, S. E., Neubauer, S. and Skinner, M.M. 2017. New fossils from Jebel Irhoud, Morocco and the pan-African origin of Homo sapiens. *Nature.* 546: 289–292.
Hume, D. 1758 [1976]. *Essays and Treatises on Several Subjects - The Natural History of Religion.* A. Wayne Cover (Ed.). Oxford:Clarendon Press.
Ingold, T. 2000. *What is an animal?* London: Unwin Hyman.
Karakostis, F. A., Hotz, G., Tourloukis, V. and Harvati, K. 2018. Evidence for precision grasping in Neandertal daily activities. *Science Advances.* 4: eaat2369.
Kochiyama, T., Naomichi Ogihara, N.,Tanabe, H. C., Kondo, O., Amano, H., Hasegawa, K., Suzuki, H., Ponce de León, M. S., Zollikofer, C. P.E., Bastir, M., Stringer, C., Sadato, N. and Akazawa, T. 2018. Reconstructing the Neanderthal Brain using Computational Anatomy. *Scientific Reports.*8(1): 6296. doi: 10.1038/s41598-018-24331-0
Kosslyn, S. M., Ganis, G. and Thompson, W. L. 2001 Neural Foundation of Imagery. *Nature Reviews Neuroscience.* 2: 635-642.
Kosslyn, S. M. and Shin, L. M. 1994. Visual Mental Images in the Brain: Current Issues. In, M. J. Farah and G. Ratcliff (Eds.), *The Neuropsychology of High Level Vision. Collected Tutorial Essays (*pp. 269-296). Hillsdale, NJ: Lawrence Erlbaum Associates.
Kouide, S. and Dehaene, S. 2007. Levels of processing during non-conscious perception: a critical review of visual masking. *Philosophical Transactions of the Royal Society of London B.* 362 (1481): 857–875.
Kozbelt, A. 2001. Artists as experts in visual cognition. *Visual Cognition.* 8 (6):705–723.
Kozbelt, A. and Seeley, W. P. 2007. Integrating art historical, psychological, and neuroscientific explanations of artists' advantages in drawing and perception. *Psychology of Aesthetics, Creativity, and the Arts* 1 (2): 80 –90.

Kreiman, G., Koch, C. and Fried, I. 2000. Imagery neurons in the human brain. *Nature.* 408: 357-361.

Laing, L. and Laing, J. 1982. *The Origins of Britain.* London: Granada.

Lakoff, G. 1987. *Women, fire and dangerous things: What categories reveal about the mind.* Chicago: University of Chicago Press.

Leakey, M. 1971. *Olduvai Gorge.* 13(84):269. Cambridge: Cambridge University Press.

LeDoux, J. E. 1994. Emotion, Memory and the Brain. *Scientific American* (June):32-39.

Leslie, A. M. 1987. Pretense and Representation: The Origins of "Theory of Mind". *Psychological Review.* 94(4): 412-426.

—. 1994. Pretending and believing: Issues in the theory of ToMM. *Cognition* (50): 211–238.

Levy, C. K. 1999. *Evolutionary Wars.* New York: W. H. Freeman.

Liu, J., Li, J., Feng,L., Li, L., Tian, J. and Lee, K. 2014. Seeing Jesus in toast: Neural and behavioral correlates of face pareidolia. *Cortex* 53: 60-77.

Lorblanchet, M. 2007. The Horse in the Palaeolithic Parietal Art of the Quercy: Outline of a Stylistic Study. In, P. Pettitt., P. Bahn and S. Ripoll (Ed.), *Palaeolithic Cave Art at Creswell Crags in European Context* (pp.207-228). Oxford: Oxford University Press.

Lorblanchet, M. and P. Bahn. 2017. *The First Artists: In Search of the World's Oldest Art.* London: Thames and Hudson.

Ludmer, R., Dudai. Y. and Rubin. N. 2011. Uncovering Camouflage: Amygdala Activation Predicts Long-Term Memory of Induced Perceptual Insight. *Neuron.* 69: 1002–1014.

Mahon, B. Z., Anzellotti, S., Schwarzbach, J., Zampini, M. and Caramazza, A. 2009. Category-Specific Organization in the Human Brain Does Not Require Visual Experience. *Neuron.* 63 (3): 397–405.

Mahon, B. Z. and Caramazza, A. 2009. Concepts and Categories: A Cognitive Neuropsychological Perspective. *Annual Review of Psychology*. 60: 27–51.

Manaseryan, N. 2003. Hunting tackle, animals and objects in rock engraving from Armenia. *Archaeofauna.* 12:193-201.

Mareschal, D. and Quinn, P. C. 2001. Categorization in infancy. *Trends in Cognitive Sciences*. 5 (10): 443-450.

Marquet, J. C. and Lorblanchet, M. 2003. A Neanderthal Face? The proto-figurine from Le Roche-Cotard (Indre-et-Loire, France). *Antiquity.* 77 (298): 661-670.

Marquet, J-C., Lorblanchet, M., Oberlin, C., Thamo-Bozso, E. and Thierry Aubre, T. 2016. Nouvelle datation du “masque” de La Roche-Cotard (Langeais, Indre-et-Loire) *Paleo*. 27: 253-263.
Nelson, K. 1977. First Steps in Language Acquisition. *Journal of the American Academy of Child Psychiatry*. 16(4): 563-583.
Öhman, A. 1986. Face the Beast and Fear the Face: Animal and Social Fears as Prototypes for Evolutionary Analyses of Emotion. *Psychophysiology*. 23 (2):123-145.
Öhman, A., Erikkson, A. and Olofsson, C. 1975a. One-trial learning and superior resistance to extinction of autonomic responses conditioned to potentially phobic stimuli. *Journal of Comparative and Physiological Psychology*. 88: 619-627.
Öhman, A., Erixson, G. and Löfberg, L. 1975b. Phobias and preparedness: phobic and neutral pictures as conditioned stimuli for human autonomic responses. *Journal of Abnormal Psychology*. 84: 41-45.
Öhman, A. and Soares, J. F. 1993. On the Automatic Nature of Phobic Fear: Conditioned Electrodermal Responses to Masked Fear-Relevant Stimuli. *Journal of Abnormal Psychology*. 102 (1):121-132.
Pavio, A. 1971. *Imagery and Visual Processes*. New York: Holt, Rinehart and Winston.
Petchkovsky, L. 2008. Some Preliminary Reflections on the Biological Substrate of Meaning Making. In, A. Dowd., C. San Roque and L. Petchkovsky (Eds.), *The Uses of Subjective Experience: A Weekend of Conversations between ANZSJA Analysts and Academics who Work with Jung's Ideas*. (pp.243-252). Proceedings of the Conference (October 20-21, 2007, e-book).
Peterson, M. A. Cacciamani, L., Mojica, A. J. and Sanguinetti, J. L. 2012. Meaning can be Accessed for the Ground Side of a Figure. *Gestalt Theory*. 34 (3/4): 297-314.
Pettitt, P., Castillejo, A. M., Arias, P., Peredo, R. O. and Harrison, R. 2014. New views on old hands: the context of stencils in El Castillo and La Garma caves (Cantabria, Spain). *Antiquity*. 88: 47-63.
Quinn, P. C., Eimas P. D. and Rosenkrantz, S. L. 1993. Evidence for representations of perceptually similar natural categories by 3- month-old and 4-month-old infants. *Perception*. 22:463-475.
Riekki, T., Lindeman, M., Aleneff, M., Halme, A. and Nuortimo, A. 2013. Paranormal and Religious Believers are more Prone to Illusory Face Perception than Skeptics and Non-believers. *Applied Cognitive Psychology*. 27: 150–155.
Rorschach, H. 1942 (4th edition). *Psychodiagnostics—a diagnostic test based on perception*. Berne:Hubert.

Sayim, B. and Cavanagh, P. 2011.What Line Drawings Reveal About the Visual Brain. *Frontiers of Human Neuroscience.* 5: 118.

Schel, A. M., Townsend, S. W., Machanda, Z., Zuberbühler. K. and Slocombe. K. E. 2013. Chimpanzee Alarm Call Production Meets Key Criteria for Intentionality. *PLoSONE.* 8 (10): e76674. doi:10.1371/journal.pone.0076674.

Schoch, W. H., Bigga, G., Böhner, U., Richter, P. and Terberger. T. 2015. New Insights on the Wooden Weapons from the Paleolithic Site of Schöningen. *Journal of Human Evolution*. 89 (12): 214–25.

Schultz, G. and Melzack, R. 1991. The Charles Bonnet Syndrome: "phantom visual images." *Perception.* 20: 809-825.

Seligman, M. 1972. *Biological Boundaries of Learning.* New York: Appleton-Century-Crofts.

Shanon, B. 2002 *The Antipodes of the Mind—Charting the Phenomenology of the Ayahuasca Experience.* Oxford: Oxford University Press.

—. 2003. Hallucinations. *Journal of Consciousness Studies.* 10 (2):3-31.

Shapin, S. 1996. *The Scientific Revolution.* Chicago: University of Chicago Press.

Shepard, R. N. 1990. *Mind Sights. Original Visual Illusions, Ambiguities, and Other Anomalies.* New York: W. H. Freeman.

Shepard, P. 1997. *The Others - How Animals Made Us Human.* Island Press: Washington DC.

Shipman, P. 2010. The Animal Connection and Human Evolution. *Current Anthropology.* 51(4):519-538.

Siegel, R. K. 1980. Hallucinations. Mind and Behavior. In, Atkinson and Atkinson (Eds.). *Readings from Scientific American* (pp. 125-132). San Francisco: W.H. Freeman and Company.

Slade, P. D. and Bentall, R. P. 1988. *Sensory Deception, a scientific analysis of hallucinations.* Baltimore: John Hopkins University Press.

Snodgrass, J. G. and Feenan, K. 1990. Priming effects in picture fragment completion: Support for perceptual closure hypothesis. *Journal of Experimental Psychology:General*. 119: 276-298.

Srinivas, K. 1993. Perceptual Specificity in non-verbal priming. *Journal of Experimental Psychology: Learning, Memory and Cognition.* 19: (3):582-602.

Taylor, R.P., Martin, T. P., Montgomery, R.D., Smith, J.H., Micolich, A.P., Boydston, C., Scannell, B.C., Fairbanks, M. S. and Spehar, B. 2017. Seeing shapes in seemingly random spatial patterns: Fractal analysis of Rorschach inkblots. *PLoSONE* 12(2): e0171289. https://doi.org/10.1371/journal.pone.0171289.

Taylor, R. and Spehar, B. 2016. Fractal Fluency: An Intimate Relationship between the Brain and Processing of Fractal Stimuli. In, A. Di Ieva (Ed.), *The Fractal Geometry of the Brain* (pp. 485-496). New York, NY: Springer.

Thackeray, J. F. 2005. The wounded roan: a contribution to the relation of hunting and trance in southern African rock art. *Antiquity* 79: 5–18.

Thieme, H. 1997. Lower Palaeolithic Hunting Spears from Germany. *Nature.* 385 (6619): 807–810.

Universität Tübingen Press Office. 2017. Successful Dig Reveals an Almost Complete Skull of a Sabre-toothed Cat. Press Release 12.04.2017.

Vyshedskiy, A. 2014 (2nd edit). On the Origin of the Human Mind. *MobileReference.* http://mobilereference.com/mind/online/index.htm.

Walther, D. B., Chai, B., Caddigan, E., Beck, D. M. and Fei-Fei, L. 2011. Simple line drawings suffice for functional MRI decoding of natural scene categories. *Proceedings of the Notational Academy of Sciences USA.*108(23): 9661–9666.

Washburn, D. K. 2018. Cognitive archaeology: a symmetry/symmetry-breaking model for the analysis of societies. *Time and Mind.* 11:2, 121-161.

Weiskrantz, L. 1996. Blindsight Revisted. *Current Opinion in Neurobiology.* 6 (2): 215-220.

White, R. 2003. *Prehistoric Art - the symbolic journey of humankind.* New York: Abrams.

Winner, E. 1982. *Invented Worlds—The Psychology of the Arts.* Cambridge, Mass: Harvard University Press.

Zhu, W., Drewes, J., Peatfield, N.A. and Melcher, D. 2016. Differential Visual Processing of Animal Images, with and without Conscious Awareness. *Frontiers in Human Neuroscience.* 10(513): 1-19.

Zilhão, J. and d'Errico, F. 2000. A Case for Neanderthal Culture. *Scientific American.* 282. (282): 84-85.

# CHAPTER SIX

# THE EMOTIONAL CONTEXT

> One cannot overemphasize the disconcerting skill with which the Palaeolithic engravers strove, from the beginnings of art, to blend their drawings with the rock, to integrate natural features into their drawings. (Lorblanchet 2007, p.214)

## Trauma, stress, emotion, hyperimagery and cave art

The model we have been developing can potentially explain the many "anomalies" of the earliest visual art and more besides. Unlike explanations based on fully-fledged hallucinatory experiences, an approach based on the psychology of perception and visual imagery can provide a more feasible explanation for the depictions that began to appear around 40,000 years ago. Despite the criticism directed towards hallucinations, this does not mean they are uninformative. We saw earlier how animals are quite common in classical hallucinations but are not solely based on cultural affinities. Shanon's detailed analysis—already alluded to—of indigenous groups and modern westerners, indicates the valuable role of hallucinations in the present context; a finding substantiated by a wide-ranging study of various psychological conditions and drug induced episodes that give rise to hallucinations, which established animals and people are the most commonly experienced items (Collerton, 2005). Shanon's hallucinations are of the full-blown variety but to understand their relevance to palaeoart it is necessary to appreciate how "normal" visual awareness can be equally informative.

Paradoxically, some scholars maintain that the perception of everyday visual reality is a type of controlled hallucination (Collerton, 2005; Clark, 2016). In one sense, this is correct because it is based on the idea that what we see is an interpretation of the visual world and not the world itself (Dror, 2005). In other words, what we see is a "representation" of visual reality that depends largely on the perceiver rather than objective reality. What, however, makes the perception of the real world compelling is that, as already outlined, it is formed out of long-standing evolutionary events (involving the practical interaction with objects such as animals and other humans that were decisive to survival). So, although we see a representation of the external world, this is nurtured by how the visual pathways were shaped by evolution—and it is one, in the normal course of events, that all humans share, thereby allowing access to an agreed, but unstated, conception of visual reality. In this connection, incoming visual information serves to regulate and update the visual system in order to confirm that the world is being interpreted correctly. In short, normal visual awareness is kept in line by visual information issuing from the real world that constrains higher-order visual imagery/memory (Collerton et al. 2005). That process is hierarchical in the sense that, at the outset, the visual image (or visual template) is built from basic geometric lines to more complex shapes culminating in the conscious identification of actual objects. To re-emphasise, this is not a one way process, as much cross-referencing and feedback occurs, including from the highest stages back to the earliest levels. This is especially the case when we need to deal with challenging viewing situations and where detailed scrutiny of stimuli is required. It is when the balance between incoming visual information and the various pathways for "representation" is disturbed or becomes unstable that hallucinatory-like episodes tend to arise (Carhart-Harris et al. 2016; Hodgson 2006a).

Humans are a species with a particular bias for the visual world, which explains why the human temporal cortex and associated visual areas underwent enlargement in hominins compared to chimps (Rilling and Seligman, 2003). That enlargement is reflected in the earlier observation that the acuity of the human eye exceeds almost all other mammals, as well as *most* birds (Caves et al. 2018). Those areas of the brain, especially the inferotemporal cortex, are acknowledged as the place where visual imagery, recognition, and visual memory occur—that is, where visual templates reside that allow the matching of incoming visual information with previously assigned "templates" or prototypes encoded in neural networks (Shepard, 1984; Kosslyn, 1994; Wu et al. 2012). Disturbance or damage to any part of that pathway can lead to the accentuation of processing in preserved areas. There are several ways by which a disturbance may be generated, with two of the major ones involving over and under stimulation through, for example, the taking of psychoactive drugs, prolonged isolation, or brain pathology.

There is another, and for our present discussion, important means by which hallucinatory-like episodes can be induced—one ignored by most investigators—concerning states of stress, anxiety, or hyperactivity in response to crucial life-events. Before describing them in detail we need to take a closer look at how the visual system at higher levels functions when faced with anomalous experiences.

We have already established that there is a common neurophysiological mechanism underlying visual processes, whether this is perception, imagining, thinking, dreaming or visual memory (Shepard, 1984; Kosslyn, 1994; Farah, 2000), namely the perceptual/recognition system (Kirby and Kosslyn, 1992). What differentiates the various levels depends on whether visual information derives from the real world, where the emphasis is on the early to middle stages, or comes from the need to reflect on events whereupon the later stages predominate. That differentiation is consistent with the experienced visual world being a product of interactions between abstract top-down memory templates and bottom-up sensory cues, with the former more influential in subjective visual awareness. Visual mental imagery and visual perception therefore share the same elements but do not draw on identical processes. Visual experience can, accordingly, be a consequence of dynamic reciprocal interactions operating between external sensory input, internal object and scene representations and goal directed attention. This can be influenced by top-down biasing, which is facilitated by object familiarity, individual goals, expectancy or predictive encoding as well as spatial attention, which influences the speed and accuracy of recognition and attention. Where perception of the world is ambiguous, expectations and predictive inference issuing from the higher visual pathway comes increasingly into play as a coping mechanism (Pinto et al. 2015; Powers et al. 2016). As well as visual memory, imagination appears to be underwritten by the same system and one of its properties is to suggest alternatives when confronted with ambiguity, especially if this is a camouflaged or indistinct form. An imaginative faculty was crucial to hominin survival as it provided instant "suggestions" as to the identity of any unrecognised entity and is the most "detachable" from everyday reality in allowing scenarios to be played out to determine potential practical implications. Such a procedure facilitates rehearsal for dealing with real events and the unexpected. Similarly, as Kosslyn and Shin (1994) emphasised, visual imagery can function in rehearsal mode by allowing imagery to selectively facilitate perception, so that if subjects imagine a form they are more likely to detect a degraded version when presented with alternatives. A by-product of this is the ability to project content and meaning onto objects that may resemble things in the real world—an ability with which humans are particular well-endowed even at a young age (Cox, 2005) to the extent meaning or agency can be projected onto neutral objects such as moving geometric shapes (Heider and Simmel, 1944)

Due to the fact neural systems need to be finely balanced, they can suffer instability as a result of injury or intoxicants etc. The taking of psychoactive drugs, for example, serves to introduce a measure of volatility by providing an internally-derived stimulus that causes considerable disequilibrium. As a consequence, imaginative and interpretative processes can come to the forefront of awareness as various spontaneously generated stimuli emerge, distort, deform, coalesce, dissipate and re-emerge in multiple ways (Powers et al, 2016). In this scenario, the visual system endeavours to make sense of the mixed and contradictory signals issuing from both external sources and within the system itself, exacerbated by mistaking internal images for real. Within this confusion, a struggle ensues causing different "hypotheses" to be generated. Though visual information from the external world may continue to exert an influence, its effects are reduced; yet the effects of an hallucinogenic drug

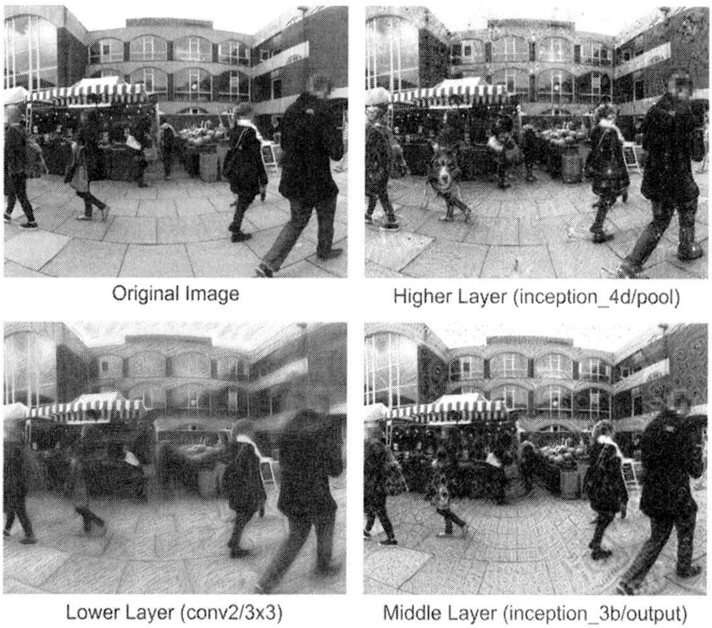


Figure 6-1. Deep Dream learning network modeling an altered state showing animals emerging from a natural scene. This begins at bottom left then to bottom right culminating in the projection of animals onto ordinary objects at top right (original in colour). (Reproduced from: Suzuki et al. 2017. Creative Commons Attribution 4.0 International).

may become so extreme full-blown "detached" visual hallucinations can occur. Such processes have recently been modelled in computer learning systems that simulate natural vision, which illustrate how a primed higher visual system becomes hyperactive to the extent illusory animals progressively seem to arise out of a normal scene (see especially top right and bottom left in Figure 6-1).

The fact humans possess a sophisticated visual aptitude that comes with an enlarged temporal cortex (and the same applies further upstream in the parietal and frontal areas), means that objects take precedence in consciousness over earlier processing levels. This is confirmed by the finding that those affected by psychoactive substances tend to be more concerned with the particular objects hallucinated (Hodgson, 2008). Nevertheless, geometric forms, which the early visual cortex prioritises, can come to conscious awareness when visual thresholds are reduced thanks to the effects of the prevailing psychoactive agent (see bottom left in Figure 6-1). Any part of the visual hierarchy is thereby liable to become available to consciousness from the earliest basic units to the later more sophisticated levels, but with a bias towards the latter (Kluver, 1994) due to the global superiority effect.

The foregoing helps explain why bright vivid colours and exaggerated forms are often experienced. These are not just a consequence of the increased sensitivity of neurons but also derive from the instability that comes when the normal markers (external sensory information) are attenuated leaving internal generated cortical events to predominate. In such circumstances, the natural tendency of imagination to disengage from visual reality can lead to even greater flights of fantasy where hallucinatory monsters, angels, hybrid figures, fabulous things and strange creatures may come to the fore.

We have described how hallucinations arise when the brain is in an abnormal state, which can inform us as to how the visual system functions when performing normal tasks. Though early researchers found that stimulation of certain areas of the visual cortex can lead to the activation of visual memories and hallucinated items (Horowitz et al. 1968; Penfield and Perot, 1963), the normal visual system is not immune to volatility as is evidenced by what is termed "The Perky Effect" (Perky, 1910). This effect shows that the division between visual imagery/memory and visual reality is fluid—a phenomenon observed when subjects were placed in a dimly lit room and asked to stare at a blank screen some distance away. The observers were asked to summon up an image of an object and while thus preoccupied (and still staring at the screen), a barely visible image of the imagined object was surreptitiously projected onto the screen. To simulate a mental image, the projected image was blurred slightly. Nearly all the observers remained oblivious to the projected image thinking that what they saw was their own mental image. As the experiment progressed, the projected image was accentuated to the extent that research personnel could easily discern the projected object, whereas the subjects remained oblivious. That phenomenon is explained by the fact that, once the imagined object became established in the mind's eye of the subjects, this masked the projected image making perception of the projection more demanding. When later informed that what they had witnessed was a real projected image, participants expressed disbelief insisting they had imagined the figure. Fascinatingly, the imagined image was often combined with the real projected object to the extent subjects reported seeing an upright image of the fruit as portrayed on the actual screen. Moreover, some observers placed the projected image within an imagined scene e.g., a leaf (real image) embellished with details (subjective image). The same effect was subsequently confirmed by other investigators (Segal and Fusella, 1970; Reeves, 1982) thereby adding weight to the notion that our hold on "visual reality" is tenuous and we are liable to be fooled or make errors in certain situations. In other words, sometimes subjective imagery can intrude on reality and vice versa. Other studies found that there is a loss of high spatial frequencies in visual imagery leading to a blurring of contours (Harvey, 1986).

Correspondingly, Neisser (1967) established that only selected parts of incoming visual information are represented in memory, which provides a starting point for the reconstruction of a past event. What is remembered about an event is also closely connected to how a person relates to that event and those aspects elaborated depend on information in long-term memory. Encoding and remembering are therefore closely aligned. As a result, ambiguity can arise, as what is remembered is constrained by what has been encoded which, in turn, depends on past experience, needs, and knowledge—all of which exert a powerful influence on what is retained (Schacter, 1996). Memory is therefore assembled from elaborations relating to how the brain imposes order on the environment in that it does not just depend on the strengths of associations but also the cues that trigger the initial memory trace. Thus, memory can be induced by selected cues from an original experience that conditions how an event is recalled. How such an event is regarded will also determine what is encoded in the memory trace (Tulving, 1983). In addition, when a retrieval cue elicits a memory, this is more likely to occur when the circumstances, especially those concerning emotional arousal, are similar. Crucially, the trigger cue combines with the original memory trace to produce an experience somewhat different from the original event. As Schacter (1996: p.71) states:

> A neural network combines information in the present environment with patterns that have been stored in the past, and the resulting mixture of the two is what the network remembers. The same conclusion applies to people. When we remember, we complete a pattern with the best match available in memory; we do not shine a spotlight on a stored picture.

As already indicated, what is remembered from an emotional experience may be influenced by arousal levels, which serve to focus attention on specific aspects of an episode. In such cases, research has found that people exposed to traumatic events tend to focus on the central themes rather than peripheral details—in other words, attention is directed towards the salient arousal-provoking aspects of the original event, referred to as "weapons focusing". For example, witnesses to a crime involving a weapon, retain accurate memories of the actual firearm but retain a distorted and incomplete memory of other aspects of the event, including the face of the person brandishing the weapon. The emotionally salient information therefore attracts attention to the extent that the broader context of the scene is often absent from memory. That effect is particularly noticeable when anxiety is high—a finding that correlates with research showing that raised levels of anxiety can lead to a narrowing of attention (Schacter,1996). Consequently, overwhelming emotional events are remembered the most and what is remembered tends to be whatever dominated a particular event. Such procedures are encoded in neural circuits by the strengthening and formation of dendrites along neural pathways.

Those observations can be applied to how everyday stress can influence or bias visual experience. For example, intense concentration on particular objects over a prolonged period, especially when associated with fatigue, can later give rise to intrusive "hallucinations" of the same object. To give some practical examples, fruit-pickers who spend long hours focussing on a particular fruit, later report seeing the sought-after item projected onto similar but different objects. Correspondingly, they might see, in their mind's eye just before sleep, intrusive images of the object concerned (often referred to as hypnagogic images). The same phenomenon occurs with biology students, who spend long hours examining cells through microscopes, where amoeba-like forms later spring to mind (Horowitz, 1968). Similarly, long-distance nighttime drivers report experiencing intrusive phantom images of headlights in more relaxed circumstances. It is as if the perceived object is so ingrained within the neural network, it can enter consciousness even in the absence of the real stimulus.

Expectations and emotional longing have also been found to induce intrusive imagery e.g., shipwrecked sailors report seeing images of boats coming to the rescue. Certain images seem to derive from purely physical causes, such as when an individual endures hunger or thirst (Gross, 1992; Sandford, 1936), where the mental image of the item in question becomes increasingly intrusive until the physical need is placated. Those kinds of "hallucinations", and others to be described below, are usually referred to as "pseudohallucinations", which take different forms (Turner, 2014), and, I would submit, are important to understanding palaeoart. However, to avoid confusion with classic hallucinations—and as delineated in earlier chapters—in the present context I refer to pseudohallucinations as projective hyperimages.[1] By way of a reminder, I mean those mental images constrained by ancestral factors that may be reinforced by ongoing experience, as was the case with UP hunter-gatherer's everyday interaction with animals.

Although in normal circumstances a memory image fades quickly, becoming less detailed, images associated with conspicuous events tend to be retained longer (Vernon, 1971). This is especially the case with stressful situations and over-arousal (Manford and Anderman, 1998). An extreme example is the shock experienced in response to a disaster or disturbing event e.g. the shell shock of soldiers during the First World War who, when later hospitalised, reported seeing the original event played out before them—sometimes referred to as "flashbacks" (Dror, 2005), otherwise known as post-traumatic stress disorder. One of the characteristics of such disorders is the reliving of the predisposing episode, which may be a coping mechanism whereby, through repetition, the "mind" attempts to familiarise itself with the event as a safeguard against the possible reoccurrence of the incident so that an appropriate level of mental defense is readily available. Horowitz, (1975) found intrusive and stimulus repetitive images in over 50% of individuals suffering mild to moderate emotional shock, thereby emphasising the widespread nature of the phenomenon. Similarly, with the death of close kin, a complete stranger might be mistaken for the deceased person (Grimby, 1998; Lange et al. 1996; Kraüpl Taylor, 1979). The stimulus driving these quasi-hallucinatory episodes is not a psychoactive agent or damage to the brain, rather the hyper-aroused limbic regions that induce imagery in the higher visual centres (Joseph, 1990). Such imagery may also be manifest in dreams or nightmares. Anything inducing over-arousal in those systems can, thereby, elicit projective hyperimagers (Manford and Anderman, 1998). The likelihood of a Perky-like effect occurring (mistaking an internally generated image for one based on external reality) is therefore greatly increased in cases of heightened arousal.

A further mental condition leading to a quasi-hallucinatory experience is when an individual suffers intense fatigue—the opposite of arousal but nevertheless a disturbed conscious state. Those affected often report seeing objects of various descriptions, invariably animals (Whitlock, 1987). For example, drivers enduring long and monotonous journeys through fog, snowstorms or at night often report seeing non-existent animals, which they swerve to avoid before the mistake is noticed (Whitlock, 1987; Mcfarland and Moore, 1957). In such cases, attention to external stimuli is compromised allowing visual imagery to come to the fore.

Lack of sleep, especially over extended periods, also leads to confusion so that one object may be mistaken for another. In extreme cases, quasi-hallucinatory episodes may occur in response to cues deriving from prevailing stimuli. For example, with sleeplessness, hyperimages may emerge in response to hints contained in the immediately perceived stimulus e.g., spots on a table might be misperceived as insects or a street sign may be mistaken for a person (Coren, 1998). Hunger, if not starvation, was probably a further predisposing factor as a lack of sustenance leads to an image of food springing to mind (Gross, 1992; Sandford, 1936). Horowitz (1975), for example, points out that any ambiguous shape can be "illusioned" into a particular object depending on the level or state of a person's primary needs.

As with full-blown hallucinations, hyperimages are experienced in subjective as opposed to external space to the extent a continuum links the two. Moreover, environmental conditions and psychological factors may increase permeability. The division between recognition, visual imagery, hyperimagery and hallucinations is, therefore, not a fixed commodity.

The diverse way hyperimages are induced provides clues as to why there are so many depictions of animals in Franco-Cantabrian caves and why they are depicted in particular ways. The behavioural phenomena already identified are highly relevant to the UP environment in which early hunter-foraging people struggled to survive. Caves are especially important for a number of reasons. However, it is unlikely UP communities ingested hallucinogenic drugs or similar and, therefore, classical hallucinations can be discounted because it is questionable that individuals resorted to such substances in the treacherous cave conditions; particularly as psychoactive agents can lead to extreme disorientation in ambiguous environments due to an overactive visual brain that is liable to misread real world cues.

So how might "hallucinations" be relevant to understanding the depictions of animals during the UP? The answer lies in projective hyperimages which, as noted, can be triggered by "normal" experiences. By exploring UP lifeways from this perspective, we are better disposed to assess their relevance to palaeoart. Fatigue, lack of sleep, various traumatic experiences, over-concentration on particular items and different kinds of stress have been identified as prime examples of events that give rise to projective hyperimages, and were regularly experienced by UP hunters. We need also to bear in mind that UP animals were considerably larger than todays with much bigger carnivores and herbivores. On hunting forays, highly focussed concentration was necessary for sustained periods in order to obtain the protein rich meat/fat diet necessary for survival. Vigilance was vital for detecting marauding predators stalking the same prey. Mental fatigue is a hallmark of focussed attention—exacerbated by the extreme physical exertion required—especially when associated with a lack of sleep. Shock and post-traumatic stress would also have been a factor as it is probable a number of palaeo-hunters were killed either by the sought-after prey, such as bison and aurochs, or by carnivores. We can imagine the trauma experienced by a hunting party seeing one of their fellow hunters trampled-on, gored, or attacked by a carnivore. The "fallen man" portrayal, a rare example of a scene in palaeoart, probably depicts such an event—albeit a bison—recast as a ritualized episode (Figure 6-2).

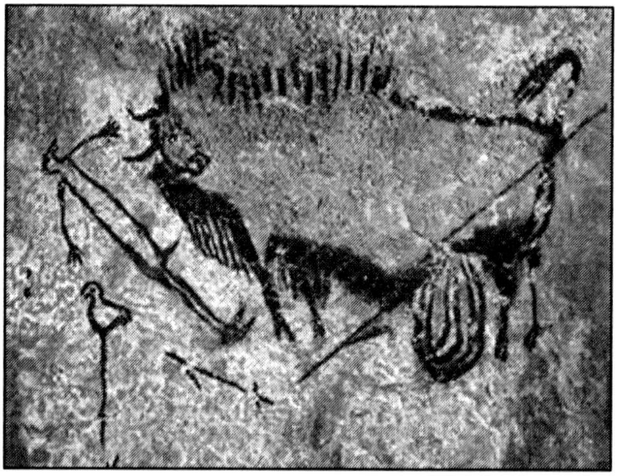

Figure 6-2. Lascaux Cave "Fallen Man" (Creative Commons Attribution-Share Alike 3.0 Unported).

In sum, mental states played a vital role in how perceptual information was processed during the UP in that the hopes, fears and expectations of hunter-gatherers affected what was perceived (Dror, 2005; Horowitz, 1975). On returning to a cave to seek safety and protection, the dim, if not dark, conditions meant that the constraining effect of external stimuli was reduced whilst, thanks to stress-related experiences, visual imagery became increasingly conspicuous. Taking into account the use of torches and lamps, this is consistent with the fact that dim light is more conducive to visual imagery than complete darkness and matches perfectly the lighting conditions of UP caves. In such an environment, various members of the group likely experienced projective hyperimages of animals thanks to any of the aforementioned reasons; further reinforced—and which has obvious parallels with the Perky Effect—by the fact many of the contours of the cave walls simulate animal outlines which, in the subdued light, led to a blurring of what separates mental images from the suggestive cave walls. Anyone who has visited the same subterranean passages will appreciate how the natural wall features suggest animal contours. But there is one more piece of vital evidence already alluded to that reinforces that effect, namely the fast implicit route already referred to for detecting animals in the emotional brain (amygdala) that occurs before conscious awareness ensues (Yang et al. 2005; Martin, 2007). Thanks to that subliminal route, in ambiguous dimly lit situations—permeated with indistinct forms resembling animals—the implicitly regulated perceptual system would automatically have primed the conscious brain to "see" animals where none existed. For example, in Figure 6-3 (Plate 1, top) one can see a natural crack that simulates the spinal contour and head of a horse, a perfect example of an implicit cue in the form of a suggestive fissure, which has been daubed with red ochre probably to mark it out as such. A little to the left of this natural cue (see bottom image in the same figure) three horses, each drawn facing the same direction as the fissure, parody the natural implicit cue, which can also be seen in the lower image to the far right. The first horse is almost an exact copy of the natural feature in that only the spinal contour and head are included. Note, that the three horses are progressively abbreviated, the relevance of which we discussed at length in the previous chapter with reference to the animals depicted in Covaciella Cave (Figure 5-22), Ekain Cave (Figure 5-23), and Le Portel Cave (Figure 5-24, top image).

One can find similar depictive strategies exploited in many other caves, such as at Rouffignac. Such priming made it almost certain hyperimages were a common experience of palaeo-people and, accordingly, provide a logical and reliable explanation for some of the strangest features of cave art.

Estimates suggest at least 10-15% of cave art incorporates natural features suggestive of animals (White, 2003).Though this is probably an underestimate as continuing discoveries imply a much greater number (Clottes, 2007), and Bahn (2016, p. 160) states that "there are countless other examples". Suggestive features of this order can be regarded as highly potent trigger cues that encouraged a visual system hyper-sensitised to perceive animals to see them even where none existed. The exploitation of natural features that simulate animals provides a fundamental clue as to the relevance of visual imagery and hyperimagery to cave art. Recall that, when information in a figure is reduced or incomplete, higher-order recognition templates become active. The natural cave features suggestive of animals exemplify such incompleteness and, therefore, automatically summon-up projective hyperimages to complete any missing parts by "filling in". To reemphasise, due to the much lower perceptual thresholds for triggering animal outlines, UP hunters would have seen more animal forms in natural cave features compared to modern humans. This is reflected in the fact that, when confronted by a threatening animal on hunting expeditions, the corresponding raised arousal levels ensured a state of vigilance so that the slightest cue suggesting a predator or potential prey evoked an image of an animal. To reiterate, that is because it is safer to assume a predator or dangerous animal is close by and respond to the slightest cue indicating a threat than to wait until a more explicit cue is forthcoming. On entering a cave or rock shelter, raised vigilance levels were necessary in order to remain in a high state of readiness so that any slight resemblance to an animal in the shifting light and shadows cast onto undulating walls was enough to evoke imagery of animals. One could consequently propose that the cave environment served as a proxy for the real hunting situation in that actual animals lay hidden in the ambiguous but suggestive caves that, again, simulated the real camouflage commonly encountered on hunting forays.

The many fissures and cracks of the cave walls incorporated into depictions corroborate the importance of projective hyperimages. The stains, blotches, concretions, shadows etc., also served as implicit trigger cues, which suggest that the influence of natural features was more widespread than commonly supposed. Yet, any surface of the cave, whether flat or undulating, provided a suitable "canvas" on which hyperimages could be projected. The flickering shadows cast by torches, fires and lamp burners also made the flat cave walls somewhat ambiguous by giving a sense of animation to the forms perceived on the various

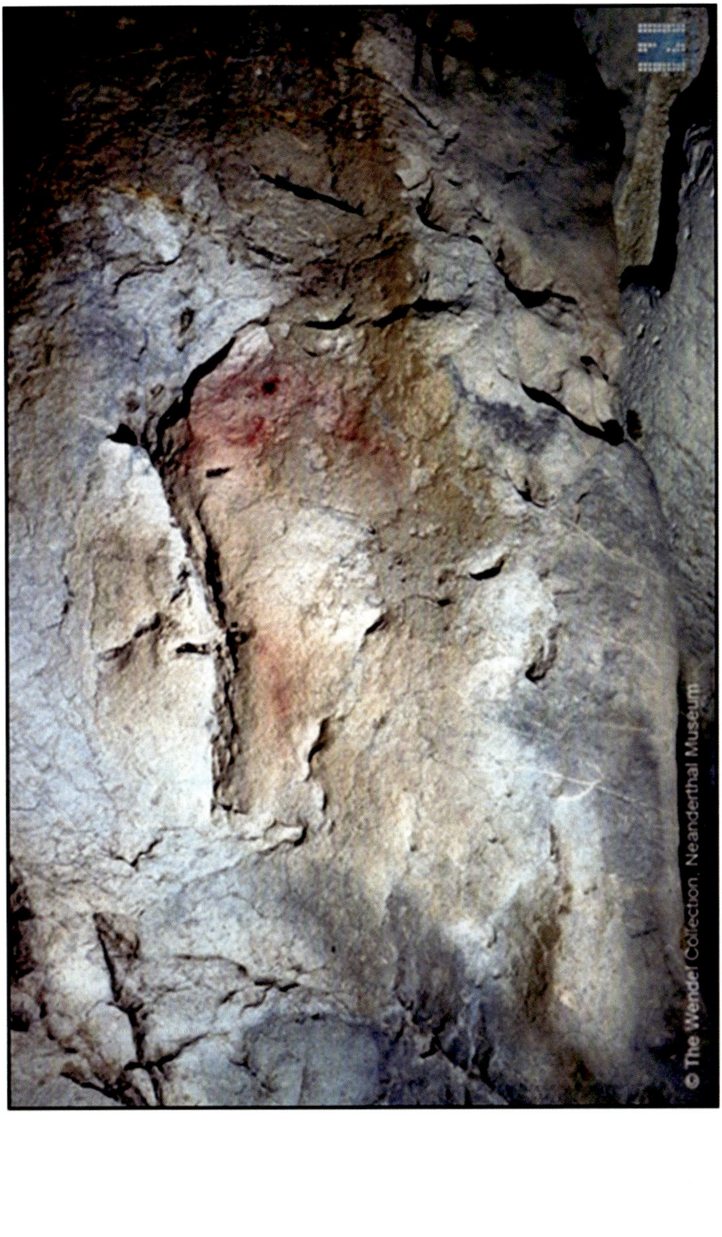
© The Wendel Collection, Neanderthal Museum

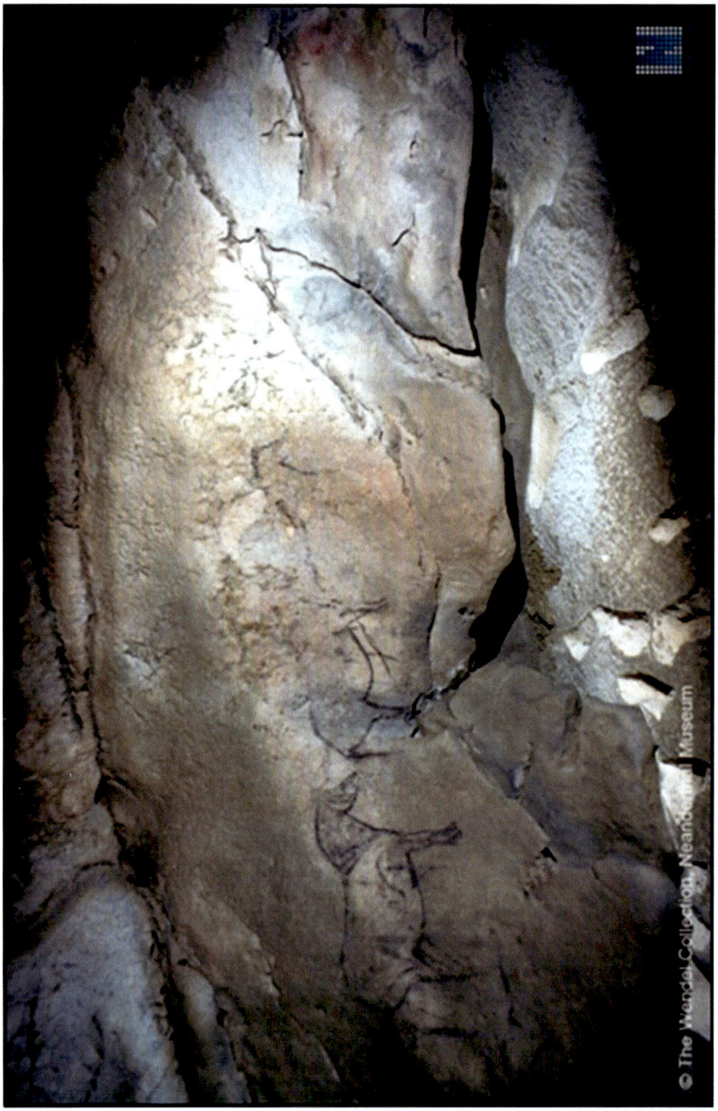


Figure 6-3 (Plate 1). Ekain Cave. The top image shows a natural wall fissure daubed with red pigment, probably to mark its resemblance to an animal. In the lower image, the natural daubed fissure can be seen to the right, and probably served as an implicit cue inspiring the creation of the three horses that display a striking similarity to the natural fissure. (© The Wendel Collection: Neanderthal Museum).

surfaces. Those experiencing hyperimages most likely remained oblivious to their derivation and, accordingly, probably regarded them as magical, thereby stimulating the need to record such "visions" in permanent form.

One of the characteristics of hyperimages (and imagery) is that they tend to be a single representative of a class (Collerton et al. 2005;Dror, 2005; Horowitz, 1968). As pointed out earlier, this is common in UP art where the outline of an animal defines its genus (Halverson, 1991). In the section on the mind's eye, we highlighted the fact that the mental image of an object is necessarily the typical or canonical view, which, in the case of animals, is the sideways profile. A finding that reinforces the fact that the animals portrayed were directly related to and derive from projected hyperimages in that only the main defining aspects of an object-class tended to be included in the depictions of UP animals.

As established, twisted perspective is fairly common and can be explained not just by the way neurons of the expedient eye function, but also by the focussing effect, as in the weapons example. Various animal parts are twisted because they trigger the typical view for crucial anatomical features that can potentially cause injury or death. A perfect example of the effect can be found in the twisted horns of the bison in the “fallen man” scene of Lascaux (Figure 6-2) where the horns are portrayed as potent weapons.

The preceding analysis explains why there are so many repeated outlines of animals in UP art, often with no regard to the position of a previously drawn figure. Having created an initial outline, the original visual image will have persisted thereby serving as the stimulus/template for a second graphic representation. The "artist", by necessity, will have stood back from the cave wall to view the first animal depiction onto which a second could then be projected. A second contour may accordingly be added slightly offset from, but overlapping with the first depicted animal. Depending on the animal in the mind's eye, a succession of outlines either of the same or different animals is likely to be produced.

The many densely packed superimpositions on cave ceilings and walls are one of the most incomprehensible features of cave art. Why should the authors spend so much time and effort creating images in such awkward places only to obliterate them with obsessional over-drawing? Again, the answer may reside in the fact they served as a prevailing record of projective hyperimages and, equally, could subsequently have acted as proxy camouflage. Furthermore, when hunters entered a cave, they will have seen hyperimages on the ceiling while standing from a number of positions, which accounts for the many caves where animals were depicted in extremely challenging locations. The bison painted on the ceiling of Altamira ceiling, which also incorporate natural features, provide just one typical example.

In contrast to superimpositions, many depictions display outlines that carefully avoid previous depictions. This suggests that, in some instances, hyperimages predominated, alternatively, existing depictions served to modulate preferences, which reflects the dynamic interaction between the suggestive natural contours and the impact of visual imagery as a function of the many variabilities existing in the caves e.g., lighting, shadows, arousal levels, expectations etc.

As noted, visual imagery is marked by the loss of high spatial frequencies causing a certain fuzziness around edges and borders (Harvey, 1986; Sekuler and Blake, 1994). Such blurriness is common in UP cave art, such as the horse panel at Ekain Cave, Spain (see Figure 5-23 above) and the bears portrayed in figure 6-4, as well as the animals depicted at

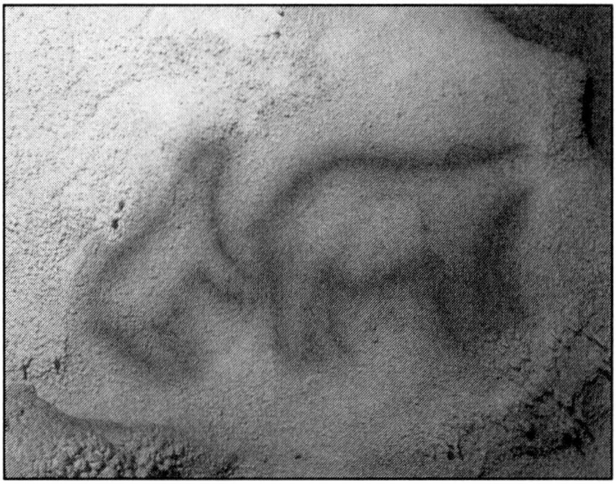

Figure 6-4. Ekain Cave, Two bears. Thanks to Jesús Altuna for permission to reproduce this image (©Jesús Altuna).

Chauvet and Asturia. Many of the depictions are incomplete or indefinite, an effect achieved either by using multiple outlines or the rubbing of depicted contours. The more distinct outlines can also be accommodated as, depending on the intensity of an experience, the image in the mind's eye can be quite distinct. This reflects the two visual streams already alluded to—the fast preconscious indistinct pathway for emotion and the higher cortical areas facilitating greater detail.

Another curious aspect of cave art is the disregard for the scale of animals portrayed in the same "scene," which parallels the fact that hyperimages are subject to size distortions to the extent that an image can range in size from the small to relatively large (Jaspers, 1963). The repetitiveness of the animal depictions also coincides with the repetitiveness of hyperimages. A further unusual characteristic is the juxtaposition of animals not normally seen together, which is all the more surprising given that some authorities claim palaeo-hunters were astutely aware of the behaviour and attributes of fauna (Clottes, 2003). The mixing of depicted fauna in such a way probably derives from intrusive images which, as well as being involuntary and unpredictable, can be elicited by a range of factors.

Focussing on animals in this way parallels the intrusive mental images of fruit pickers mentioned above. Intrusive images of animal hoof-prints could similarly have emerged in the mind's eye of cave artists leading to the depiction of suchlike. Moreover, in the white snowy environment of the Ice Age, animal tracks would have stood out against the all-pervading whiteness thereby increasing saliency. So it unsurprising that the depiction of animal tracks have been found on some of the earliest finds of palaeoart at sites such as Abri Blanchard and Abri Cellier (White et al. 2017; Hodgson, 2018) (more on hoof-prints below).

Thus, cave walls were not so much a membrane through which a shaman sought to pass in order to reach a realm populated by animal spirits as Lewis-Williams (2002) proposes, but rather a resonant surface where a hunter's mental imagery and the suggestive cave wall constantly interacted. It is therefore not what was perceived to exist beyond the rock that is crucial but rather the subtle psycho-visual interaction between the observer and the actual surface of the cave wall. In sum, the reduced light of the caves, the suggestive animal-like natural features, and the highly charged visual system of palaeo-hunters combined making it almost certain that hyperimages of animals were regularly experienced, thus motivating and directing their depiction.

The main similarities between hyperimages and Franco-Cantabrian cave art can be summarised as follows:

- Incompleteness
- Outline contour
- Blurred quality of edges-faded quality
- Size disparities
- Exaggeration and concern for salient features
- Single units
- Basic category or class
- Repetition
- Typical sideways view
- Inclusion of animals not normally seen together

The range of psychological factors eliciting hyperimages of animals can be summarised as follows:

- Prolonged concentration levels directed at one kind of object
- Fatigue
- Lack of sleep
- Expectations
- Fear and threat of attack leading to raised levels of arousal
- Hunger
- Mild to moderate shock
- Traumatic events e.g., the death of kin or fellow hunter by animals (on rare occasions).
- Hypnagogic and hypnopomic states
- Light headedness
- Raised levels of adrenaline
- Existence of an ancestral neural network for detecting animals.

The main environmental criteria facilitating hyperimages in caves are as follows:

- Darkness
- Natural features simulating animal contours
- Flickering lights from torches and fires
- The presence of dangerous predators
- Raised carbon dioxide levels
- Smoke from fires
- Shadows

Those findings allow us to gain insights into UP art in ways that would not otherwise be possible. For example, the apparent naturalism at such an early date can be explained by the fact that hyperimages were readily accessible to UP hunters. In essence, an UP individual only needed to add one or two marks to an existing suggestive rock feature to complete a naturalistic representation of an animal. The fact that the representation of animals persisted for 25,000 years in virtually the same format provides further corroboration in that the visual template sustaining this naturalism was based on hyperimages. The interaction between the suggestive cave environment, hyperimages and the daily experiences of hunter/gatherers provides a potent means to explain that longevity and consistency. Fundamentally, because human visual memory is unreliable—in that memory is not a complete snapshot of a past event but is constructed from fragments of an episode that is fused with prevailing circumstances—we would expect a massive variation in the objects depicted but this is plainly not the case.

The fact that handmarks (positive and negative) are frequently located adjacent to evocative natural cave wall features (Pettitt et al. 2014) indicates they may be related to marking areas where rock formations were psychologically evocative (see, for example, Figure 5-10 above and and 6-3). Handmarks have also been linked to shamanism, yet they predate iconic images, as at Cosquer (Clottes, 2000), El Castillo and Altamira (Pettitt et al. 2015) and, at a cave in Spain, were probably made by Neanderthals 65,000 years ago (Hoffman et al. 2018). Handmarks may derive from the fact they are a relatively easy to produce, which is corroborated by the fact children made some of the prints. This suggests handmarks emerged independently of the animal representations, though it is probable they did, eventually, come to be associated with the faunal images (more on this below). In the present context, the hypothesis is that, when specifically associated with animal outlines, handmarks were a means of testing or gauging the tangibility of animal representations in the face of the intangibility of hyperimages.

The important point has been made that graphic representations are themselves no more than geometric patterns that are constructed in such a way that they invite us to perceive them as objects (Kosslyn and Thompson, 2003). Similarly, visual imagery can be regarded as a kind of "sketch" (encodes only the more obvious contour primitives of an object such as the outline) but when greater detail is called for, the lower end of the visual cortex is recruited. As the early visual cortex (e.g., V1 and V2) is particularly concerned with processing geometrics, the depiction of such "primitives" adds what Horowitz (1975, p.178) terms a "perceptual type of intensity to the resultant constructions." The inclusion of patterns with figurative depictions may be because their inherent appeal facilitates interest due to the fact that, as distractors, they helped supplement the camouflage effect of caves.

The range of phenomena in cave art that can potentially be explained by hyperimages is summarised as follows:

- The preoccupation with animal outlines for 25,000 years
- Superimpositions
- Haphazard placement of images
- Location of depictions on ceilings and in deep recesses
- Caves as favoured location

- The inclusion of undulating surfaces and fissures as part of the depictions
- The universality of the typical viewing profile
- Size differentials of animals
- Various orientations of animals on walls and ceilings
- Indistinct, incomplete and blurred outlines
- Multiple contours
- Apparent naturalism at an early date
- Animals portrayed together that do not co-exist in reality
- Twisted perspective
- Hooves twisted towards the viewer
- Hybrid figures
- Geometric motifs
- Hand stencils/prints
- Repetition of the cervico-dorsal contour of animals

Those factors will have been underwritten by the fact that animals were integral to the evolution of the human brain to the extent that, as we have seen, animals have a dedicated domain within the temporal cortex. This is supported by the fact that studies of eye movements when viewing complex scenes found that attention gravitates more to people and animals than inanimate objects (Yarbus, 1967). Moreover, we saw that one of the commonest objects seen by those experiencing hallucinations is animals (Collerton et al. 2005; Manford and Anderman, 1998). This is corroborated by the fact that, even in today's highly technological societies, large predatory cats, such as pumas, are often mistakenly seen prowling around urban environments and across farmland, even though the evidence for such felines is non-existent. Such occurrences may derive from hyperimages primed by rumours and media reports often accompanied by blurred or incomplete photographs of the supposed creature. Correspondingly, Revonsou (2000) found modern humans continue to dream of being attacked by animals that may stem from our ancestors' long-term interaction with fauna. Franklin and Zyphur (2005) additionally suggest dreaming provides a form of rehearsal for assisting survival in real life situations. This chimes with the earlier observation that the depiction of animals served as an aide to visual imagery and memory in promoting survival by helping prime perception for detecting animals with greater efficiency. Interestingly, research demonstrates that for organisms, such as spiders and snakes, areas of the human perceptual brain become active even when individuals are not consciously aware of their presence (de Jesus Gomes, 2016), which complements similar research on the perception of animals (Mormann et al. 2007; Zhu et al. 2016). Even more interesting, Mormann et al (2007) found that the right side of the human amygdala is sensitive not just to predators but also benign animals; a finding that reflects the approach/avoidance scenario experienced by hunters with regard to carnivores and prey, respectively.

Taken together with the fact animals were crucial to the everyday lives of ancient hominins, it is hardly surprising that the profiles of fauna should come to mind so often and were so pervasive in palaeoart. This is consistent with the idea that the behavioural relevance of an item biases attention and explains why visual imagery mirrors expectations (Collerton et al. 2005; Henderson and Hollingworth, 1999). As stipulated, implicit perception is highly relevant in that recent and repeated visual exposure to an object has a tacit (subconscious) influence on visual memory recall (Hodgson, 2003). In addition, parts of a scene that are the subject of focused attention tend to become exaggerated in both hyperimages and hallucinations (Henderson et al. 2001), which is reflected in the fact exaggerated features are widespread in UP art.

The finding that imagery is crucial for identifying objects in visually "noisy" environments (Kirby and Kosslyn, 1992) is especially important because the many superimposed outlines of animals in caves not only stimulate mental imagery but also lead to positive feedback that inspire the depiction of further animals.

One aspect of UP art not yet discussed to any great extent is portable art, which refers to depictions engraved on blocks and weapons such as spear throwers, but also includes figurines. Again, such objects can be explained by the intimate relationship with animals. That relationship, it should be remembered, existed since—or even before—our early ancestors descended from the trees some three million years ago. Such a concern for animals is a worldwide phenomenon reflected in the way animals are portrayed in a similar format to that found on European manuports (Hodgson and Watson, 2015). It is particularly notable that many of the animals depicted on manuports are weapons employed to hunt animals, which suggests that animal portrayals were utilised to increase the potency of such objects. Around the time of the first parietal art in Europe, statuettes were being produced in south-west Germany consisting of animals, such as bears and lions, many which display threat and power, often in a pre-attack posture (Hahn, 1986). This is substantiated by the fact that many animal figurines derive from the natural shape of the stones from which they were made, of which some almost completely derive from the natural form with only one or two intentional hints, as many of the statuettes from Bédeilhac Cave in France testify. Some of those figures are referred to as "tête d'angle" precisely because they follow the natural contour of the stone or rock. On many, it is not just the contour that is incorporated but also protrusions and cavities, just as in the caves (Sauvet, 2004). Equally, an exhaustive study of stone plaquettes containing engraved animal outlines, found that the natural rock shape was critical not only to the initial depiction but also to subsequent images as part of the ongoing, repeated use and modulation of the blocks (Needham, 2017). As Needham states:

> The material selected for the manufacture of plaquettes can be understood to have been active in the process of creating the engraved form, through the incorporation of evocative details in its shape and surface morphology. (Needham, 2017. p.199).

As with cave depictions, scholars claim that the incorporation of natural features in sculptures has been overlooked (Clottes, 2007). In fact, the natural shapes of fallen rocks seem to have acted as an implicit cue for completing animal heads and the hyoid bone of horses, which is itself suggestive of a horse's head, similarly acted as the initial inspiration for creating a depicted horse head (Sauvet, 2004; Conneller, 2011), as illustrated in Figure 6-5. The more elaborate sculptures may also have been inspired by natural features but, due to the more extensive working, the natural predisposing features are no longer visible. One exceptional example from Zaraysk in Russia, however, shows how the natural curvature of a mammoth's tusk was utilised to form the back arch of the sculpted bison (Figure 6-6). These insights suggest a close relationship between portable objects and the cave depictions in that both depended on a close interaction between the natural features and projective hyperimages.

To summarise, Franco-Cantabrian caves served as a particularly favourable location that encouraged hyperimages. Hunters entering the caves had often engaged in stressful and demanding situations on hunting forays involving highly focussed and prolonged concentration that led to an accentuation of visual imagery at the expense of externally perceived stimuli. This was reinforced by hunger or traumatic experiences relating to the dangers of tracking game and avoidance of predators whereby survival of individual hunters regularly came under threat. Hyperimages were frequently experienced during the UP as a result of hunger, fatigue, prolonged focussed attention, plus various degrees of stress. Because palaeo-hunters remained unaware as to the cause of hyperimages, they probably regarded them as special. In that way, hyperimages—especially in the dimly lit caves—were regarded as akin to "visions". The fact that the depictions of fauna show many similarities with hyperimages, especially with regard to incompleteness, outline contour, blurred quality, exaggeration of features and single units, strongly suggests that this type of imagery provided the initial inspiration for the depictions of animals.

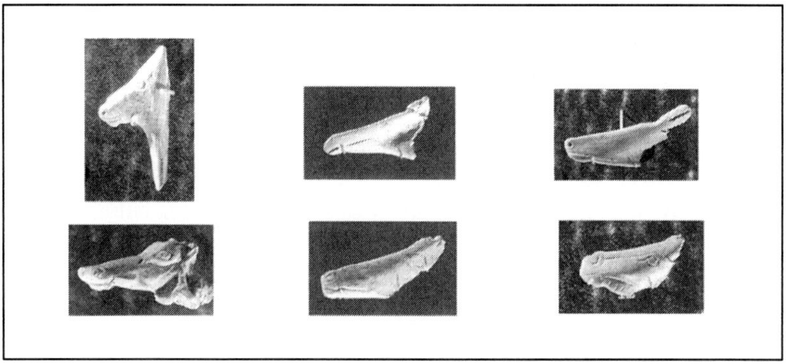

Figure 6-5. Examples of horse hyoid bones where the natural shape was used as the basis for creating the horse heads (photo by author).

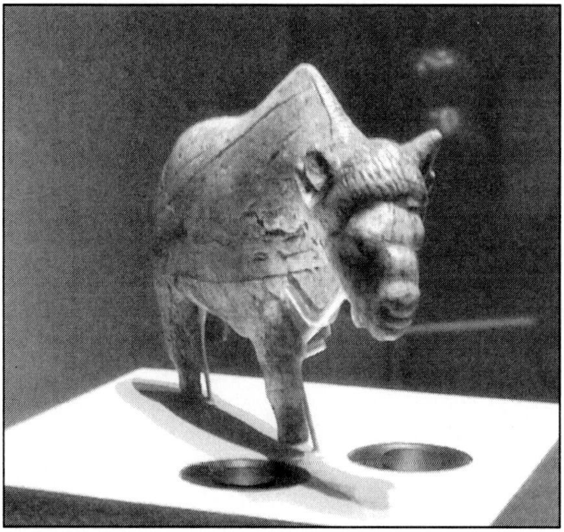

Figure 6-6. The Zaraysk bison from Russia, c.21.000 BP (in the Zaraysk Museum of Art and History). The sculpture was made from a mammoth tusk where the natural curvature was exploited to form the dip in the bison's back. (Photo by the author).

Unlike many previous explanations, the present approach is based on the everyday activities of UP people as they attempted to survive during the fluctuating climate of the Ice Age—conditions that directly influenced their psychological state which, in turn, determined the nature of the depictions. For the reasons stated, it is unlikely—as suggested by some—that the hybrid anthropomorphic figures were related to shamanism at such an early date. Rather animism, or what has recently been defined as relational epistemology (or new animism), which regards the human and animal worlds as fluid (Harvey, 2005; Bird-David, 1999), seems to be more apposite. This is borne out by the many present and past tribal groups where artistic activities are premised on an animistic belief system. Although animism may be relevant to UP, the striking uniformity in both subject matter and style over a prolonged period can only by explained by the way the visual brain functions together with a hunter-gatherer lifestyle. As already emphasised, if cultural factors were preeminent, a much greater variation in subject matter and style should be evident but this is clearly not the case. It should be remembered that in most, if not all, cultures where representational art exists, substantial differences occur in subject matter and style over relatively short periods as a result.

## The depiction of people

Although animals are by far the commonest subject in palaeoart, the second is people, which mainly represent females. The female figure is usually represented unclothed in outline, from either the front or side, but the pubic triangle in isolation also occurs regularly throughout the UP. The evidence from the famous Venus figurines and the 2D representations suggest they relate to sexual prowess and fertility, supported by the fact that the overtly sexual parts of the female body, the vulva and breast, are invariably exaggerated whereas the head (in 2D) or facial features (in sculptures) is generally ignored. In contrast, the male figure—though rare and difficult to discern—often includes the head, shoulders, some facial features (Bahn, 2016) and is represented with clothes. Male and females were, nevertheless, drawn from both the front and side (foremost and subsidiary primary views).

The concentration on particular parts of the female body suggests the stimulation of specific cortical domains that encode human attributes associated with the emotional brain. That is, the raw sign-stimuli that arouse male sexual interest, which were particularly prominent in the visual repertoire of males, leading to their appearance in palaeoart (Guthrie, 2005). At the rock shelter of Roc aux Sorciers, in Angles-sur-

Anglin, the sexual potency of the suggestive bas-relief figures of a group of female torsos is obvious (Figure 6-7), where the pubic triangle is prominent, and has been accentuated by exploiting the natural depressions of the rock surface (many of the animal carvings at the same site similarly exploit the natural rock contours). The importance of the influence of natural rock features is underlined by Le Guillou:

> The example of this Venus and the other female representations seem to show that the general morphology of the cave and the localized morphologies of the galleries and walls are not just taken into account in the creation of the parietal design: they induce it (Le Guillou, 2001, p. 3).

A Venus-like figure has also been found at Chauvet ("The Pont-d'Arc Venus") composed of no more than the legs and pubic triangle. Four other female pubic triangles were additionally discovered in the same system (Le Guillou, 2001).

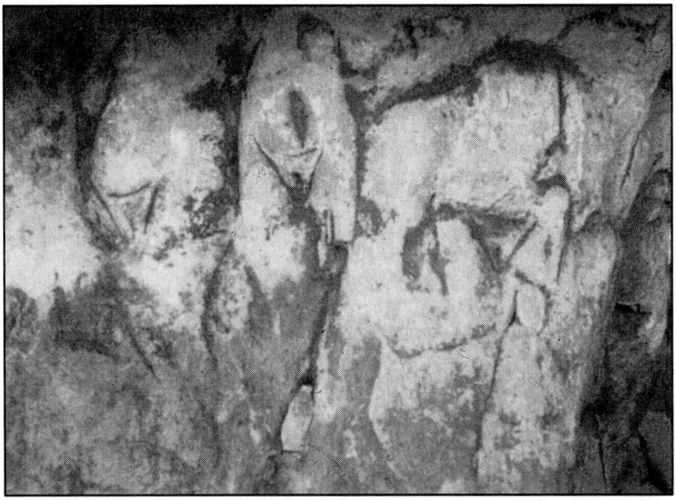

Figure 6-7. Rock Shelter of Roc aux Sorciers highlighting the female figures and pubic triangle (Photo by JYB Devot, Creative Commons Attribution-Share Alike 4.0 International).

A word of caution is in order here because many of the reported pubic triangles may actually represent animal tracks, especially as the latter display more similarities to tracks than vulvae (Hodgson, 2018). This suggests that it should not be assumed that circular or triangular motifs automatically represent vulvae unless obvious contextual information is available, namely relevant corresponding anatomical components.

Encoding of the human form in the brain is separate but adjacent to that of animals, which suggests, because of their importance in the social sphere during the evolution of *Homo*, part of the brain underwent specialisation in order to decode human facial signals and anatomical gestures. As a subset of that domain, the ability to respond to specific sexual signals was fundamental. By depicting such elemental cues, Palaeolithic people were, in effect, expressing their preeminence. That does not mean males rather than females were wholly responsible for the human images, as the same signals will have been just as important to female sexuality. So, the representation of women can be interpreted as typifying the psycho-sexual dynamic existing between males and females. Interestingly, it was almost always male bodies that were buried in graves together with ornamental artefacts (Pettitt, 2002). A rare example of an adult female burial comes from El Mirón Cave in Spain (over 15,000 BP) which, like many of the Venus figurines, was covered in ochre, and interred hidden between a slab of rock and a rear chamber wall (Straus et al. 2015). Similarly, many Venus figurines were hidden in cave recesses or were buried in pits suggesting a connection with ritual.

## Problems representing humans

A formal analysis of the depiction of people by Deregowski (1996) concluded that the scarcity of the human form in palaeoart can be explained by the peculiar difficulties of portraying the human figure. Deregowski's conclusion, however, is based on the difficulties experienced in drawing cats. Cats, Deregowski claims, have, like humans, very flexible bodies in the sense that the body shape lacks an obvious typical outline. Compared to the average animal, whose typical profile can be gauged from the spinal contour, this makes the task of assigning a typical view problematic. The evidence against this resides in the fact that, in palaeoart, cats—albeit the larger variety—are nearly always drawn in typical profile view despite the apparent difficulties. That may be because cats, when depicted, are liable to the same perceptual constraints that apply to the majority of animals, and the spinal contour does seem to be a prominent feature even in felines despite their supple anatomy. After all, we still recognise cats as cats, so the same perceptual/recognition principles apply.

Deregowski makes a more substantive case for the lack of humans and their relative simplicity in depictions based on the observation that the human body is a complex arrangement of various planes compared to the straightforward bi-symmetry of most animals, thus making it harder to determine the overall profile of the human body. This reminds one of the difficulties in trying to draw creatures, such as crocodiles, which also lack a strightforward bi-symmetrical shape. However, (Schmid, 1984) has shown that we should expect to find both the frontal and side view of humans represented in 2D depictions because they are typical features of the human body, and that is exactly what previals in UP art.

Despite the rarity of human figures in palaeoart, increasing numbers are being found that were previously overlooked due to the fact they were located on the ground rather than walls or ceilings (Bahn, 2016).[2] Because of this, many were obliterated by the footprints of investigators, who did not expect to find suchlike on cave floors and are now lost. The depictions of humans may thus be much more numerous than previously assumed and their rarity can, to some extent, be accounted for by a quirk of archaeological investigation. In fact, the few surviving depictions of human faces (of which a few may be females), although unclear, show a concern for detail by displaying facial expressions, beards, hairstyle or a particular type of head covering. A finding that, again, contradicts Deregowski's notion palaeo-artists found humans difficult to portray and, for that reason, are rare. Nevertheless, the male figure is often poorly displayed and difficult to single out. However, the general lack of faces and their poor representation is surprising given that it is such an important factor in human communication with a dedicated area of the cortex devoted to its discrimination (Barton, 2008), especially as faces are common in later rock art (Watson, 2008; Valle et al. 2018). This may be due to the fact that, as hunters, palaeo-humans were more concerned with animals in that their visual system was primed to perceive fauna in caves as well as at other sites.

One other factor relevant to the poorly represented human images relates to the possibility that UP people did not see themselves as separate from animals but rather regarded them as "other than human persons" to the extent that human presence was amalgamated with the animal depictions in a number of ways (Hodgson, 2017). For example, many of the facial expressions of the animals depicted portray human-like characteristics.

Where does this leave us in terms of how humans are depicted? The main point I wish to emphasise is that the female figure was probably based on key-stimuli appertaining to psycho-sexual imperatives. The depiction of males, however, seems to transcend such factors with an emphasis on the individual rather than the prototype. Despite the ability to depict individuals, the formal graphic characteristics centred on typical views, as humans were drawn from the front 90% of the time (Bednarik, 2003) and the body from the side i.e., the foremost and one kind of subsidiary view, respectively—both of which correspond to the views proposed by Schmid (1984).

## Eyes in hand

One feature of human anatomy commonly depicted in rock art sites throughout the world is the hand (Walker et al. 2016). We saw earlier that this may be because the hand can easily be employed to make a handmark. One handmark has even been dated to a minimum of 64,000 BP in Spain (e.g. Hoffmann et al. 2018) and appears to be the oldest example of figurative art, predating the depictions of animals in Palaeolithic Europe by several thousand years and may, therefore, have been made by a Neanderthal. Similarly in Australasia, at Sulawesi, (Indonesia), a handmark has been dated to at least 39,900 BP. In the light of such discoveries the chronology of handmarks (García-Diez et al. 2013; Pettitt et al. 2015) has been revised in that many are deemed to be far more ancient than originally assumed. As a result, figurative art depicting animals turns out to be a more recent phenomenon than the most ancient handmarks at UP sites across Europe. Similarly, a babirusa ("pig-deer") from Sulawesi (Aubert et al. 2014) has been dated to 35,400 BP, which again is a few thousand years later than the dated handmark from the same site. A handmark from a cave in Borneo may also be 52,000 years old, with the large wild cattle depicted in the same cave (cited in the previous chapter [Aubert et al, 2018]) dating to around 40,000 years ago. Correspondingly, in the rock art of Arnhem Land, Australia—long thought to be relatively ancient—handmarks predate the animal depictions (Chippindale and Taçon, 2000).

Handmarks, therefore, stand as perhaps the most ancient and universal type of figurative depiction precisely because they are a relatively straightforward and accessible means of signaling human presence (Hodgson, 2006b; Walker et al. 2016), and which are easily produced in close proximity to the body—but this begs the question of how handmarks first came about.

Everyday activities may have left accidental handmarks on natural surfaces. For example, the butchery of animals would have led to hands covered in blood giving rise to accidentally made handprints on various surfaces (Hodgson, 2006b, 2013). When ochre came to be used—for which there is evidence in Europe by 250,000 BP (Roebroeks et al. 2012) and Africa by 300,000 BP (Barham, 2002)—handmarks could have been produced accidentally in making and using pigment, perhaps for the purpose of body colouration. In caves, the act of navigation in dark spaces can leave hand, and foot, impressions on surfaces and handprints from the UP still survive in cave earth and mondmilch or soft clay (Clottes et al. 2005; Pastoors et al. 2015). In caves, making handmarks probably began with hands touching suggestive natural features during the exploration of such formidable places, which is supported by the fact that many handmarks are located on or near specific natural features rather than placed randomly on available surfaces (Pettitt et al. 2014). Thus, handmarks probably originated from accidental, messy activities, which subsequently became associated with ochre, perhaps in the context of marking of the body. This can potentially explain the origin of positive hand prints but the majority of handmarks in Palaeolithic art take the form of negative stencils. Conceptually, negative stencils require greater planning and care than simply pressing a pigment-covered hand onto a surface. Yet, progressing from the relatively simple handmarks to the first animal depictions is an even more challenging cognitive leap. Nonetheless, the "trick" of depicting a hand may have implanted in the Palaeolithic mind the idea that one thing could stand for another and, ultimately, that animal depictions were possible (Hodgson and Pettitt, 2018). Intriguingly, the fact that the first handmarks in Europe were made around 64,000 BP means that Neanderthals may well have "handed on" to later humans the idea that representational iconicity was possible. One interesting observation regarding handprints is that many clearly display five fingers, though some appear to display bent digits. It seems that this configuration provided the most obvious way of achieving an unequivocal depiction (Walker et al. 2016), especially when it is realized that the hand is particularly malleable in that it can be flexed in numerous ways. In effect, the hand became a means of representing human presence in caves as well as at numerous sites globally. The prevalence of handmarks may account for the lack of the actual depiction of faces and the human form in that handmarks acted as a potent signature of human presence (Hodgson, 2013). Moreover, as in Gargas cave and elsewhere, some early animal outlines were made in mondmilch using fingers. In both cases, there is a strong element of tactility (or palpation) in the placement of the handmarks on the natural surfaces (Pettitt et al. 2014). However, even though handmarks demonstrate that iconicity was possible, a further device was required to facilitate the animal depictions. As indicated, this took the form of the hunter's keen eye for detecting animals and the projective imagery that came to the fore within caves containing suggestive walls and surfaces. Semiotically, that procedure probably began with the indexical (hand representing complete human), then specific animal features (also indexical) representing the whole animal, ultimately culminating in full iconicity with the inclusion of many anatomical details (Hodgson, 2013; Froese, 2017). Given that handmarks predate animal depictions by a considerable margin on several continents, this suggests they served as an important precursor to the faunal images. Yet, handmarks were probably not the only precursor that gave rise to the idea that the depiction of animals was possible.

We saw in a previous chapter that a crucial antecedent to animal depictions may have been the ability of hunters to exploit disguises by wearing animal skins in order to approach fauna. Such strategies may have been so convincing that, from a distance, humans so disguised could have been mistaken for animals by fellow hunters (Hodgson and Helvenston, 2006a), especially as clothing was made from animal skins and furs, that was especially effective for staving off the cold during the Ice Age UP Europe. We also need to bear in mind that hunters are supremely skilled at identifying and interpreting the behavior of animals from tracks (Avital, 1998; Hodgson and Helvenston, 2006b). In a way, they were already tuned to the notion that an abstract category, i.e., a track/hoofprint, represented an animal in that they could "see" the animal in its tracks.

Those insights suggest a number of factors interacted to prime the ability to depict animals. Although handmarks may have acted as an important bridge towards realising the more complex animal depictions (Hodgson, 2006b, 2013; Malotki, 2013: Pettitt et al. 2015), the ability to enact animal disguises and decipher animal tracks was required before this could occur.

## Notes

1. Despite deriving from and having similarities with pseudohallucinations, hyperimages have important differences in that they are primed by long term evolutionary factors relating to underlying cognitive domains.

2. Depictions of faces, some on the ground, exist in the cave of Le Marche in southern France. Dr Michael Rappenglueck, of Munich University, believes that many more of these faces have been overlooked because they were regarded as fakes (see Whitehouse, 2002).

## References

Aubert, M., Brumm. A., Ramli, M., Sutikna, T., Saptomo, E. W., Hakim, B., Morwood, M. J., van den Bergh, G. D., Kinsley, L. and Dosseto, A. 2014. Pleistocene cave art from Sulawesi Indonesia. *Nature,* 223–227.

Avital, T. 1998. Footprints Literacy: The Origins of Art and Prelude to Science. Symmetry: Art and Science. *The Quarterly of the International Society for the Interdisciplinary Study of Symmetry*. 9 (1): 3–46.

Bahn, P. G. 2016. *Images of the Ice Age.* Oxford: Oxford University Press.

Barham, L. S. 2002. Systematic pigment use in the Middle Pleistocene of South-Central Africa. *Current Anthropology.* (43)1: 181–190.

Barton, J.J. 2008 Structure and function in acquired prosopagnosia: lessons from a series of 10 patients with brain damage. *Journal of Neuropsychology*. 2: 197–225.

Bednarik, R. G. 2003. A Figurine from the African Acheulian. *Current Anthropology.* 44 (3):405-413.

Bird-David, N. 1999. "Animism" Revisited. Personhood, Environment, and Relational Epistemology. *Current Anthropology.* 40: 67-91.

Carhart-Harris, R. L., Muthukumaraswamy, S., Roseman, L., Kaelen, M., Droog, W., Murphy, K., Tagliazucchi, E., Schenberg, E.E., Nest, T., Orban, C., Leech, R., Williams, L.T., Williams, T.M., Bolstridge, M., Sessa, B., McGonigle, J., Sereno, M. I., Nichols, D., Hellyer, P.J., Hobden, P., Evans, J., Singh, K.D., Wise, R.G., Curran, H.V., Feilding, A. and Nutt, D.J. 2016. Neural correlates of the LSD experience revealed by multimodal neuroimaging. *Proceedings of the National Academy of Sciences of the United States of America.* 113 (17): 4853-4858.

Caves, E. M., Brandley, N. C. and Johnsen, S. 2018.Visual Acuity and the Evolution of Signals. *Trends in Ecology & Evolution*. 33 (5):358-372.

Chippindale, C. and Taçon, P.S.C. 2000. The many ways of dating Arnhem Land rock-art, north Australia. In, C. Chippindale and P.S.C. Taçon (Eds.), *The Archaeology of Rock Art* (pp. 90–111).Cambridge: Cambridge University Press.

Clark, A. 2016. *Surfing Uncertainty: Prediction, Action, and the Embodied Mind.* Oxford University Press: Oxford.

Clottes, J. 2000. The "Three Cs": fresh avenues towards European Palaeolithic art. In, C. Chippindale and P. S. C. Taçon (Eds), *The Archaeology of Rock Art* (pp. 112–29).Cambridge: Cambridge University Press.

—. Return to Chauvet Cave. *Excavating the Birthplace of Art: The First Full Report.* London: Thames and Hudson.

—. 2007. New discoveries at Niaux Cave in the French Pyrenees. In, R. G. Bednarik and P.C. Reddy (Ed.), *Exploring the Mind of Ancient Man: Festschrift to Robert, G. Bednarik* (281–91). New Delhi: Research India Press.

Clottes, J., Courtin, J. and Vanrell. L. 2005. Prehistoric images and medicines under the sea. *International Newsletter on Rock Art (*42): 1-8.

Collerton, D., Perry, E. and Mckeith, I. 2005. Why people see things that are not there: A novel Perception and Attention Deficit model for recurrent complex visual hallucinations. *Behaviorial and Brian Sciences.* 28: 737-794.

Conneller, C. 2011. *An Archaeology of Material: Substantial Transformations in Early Prehistoric Europe.* London: Routledge.

Coren, S. 1998. Sleep Deprivation, Psychosis and Mental Efficiency. XV(3), *Psychiatric Times.* http://www.psychiatrictimes.com/p980301b.html

Cox, M. V. 2005. *The Pictorial World of the Child.* Cambridge: Cambridge University Press.

de Jesus Gomes, N. M. 2016. The Advantage of Fear Stimuli in Accessing Visual Awareness. Dissertation presented to the Universidade Católica Portuguesa for of the master's degree in Neuropsychology. https://repositorio.ucp.pt/bitstream/10400.14/.../NG_Dissertação_Mestrado_Final.pdf

Deregowski, J. B. 1996. A man is a difficult beast to draw: the neglected determinant in rock art. *Aesthetics symposium (IFRAO),* Swakopmund, Namibia. http://cogweb.ucla.edu/ep/Art/Deregowski_96.pdf

Dror, H. E. 2005. Perception is far from perfection: The role of the brain and mind in constructing realities. *Behaviorial and Brian Sciences.* 28: 763.

Farah, M. J. 2000. *The Cognitive Neuroscience of Vision.* Malden: Mass.: Blackwell.

Franklin, M.S. and Zyphur, M.J. 2005. The role of dreams in the evolution of the human mind. *Evolutionary Psychology.* 3:59–78.

Froese, T. 2017. Making sense of the chronology of Paleolithic cave painting from the perspective of material engagement theory. *Phenomenology and the Cognitive Sciences.* 1-22. https://doi.org/10.1007/s11097-017-9537-8

García-Diez, M., Hoffmann, D., Zilhão, J., de las Heras, C., Lasheras, J. A., Montes. R. and Pike. A. 2013. Uranium-series dating reveals a long sequence of rock art at Altamira Cave (Santillana del Mar, Cantabria). *Journal of Archaeological Science.* 40 (11): 4098–4106.

Grimby, A. 1998. Hallucinations following the loss of a spouse: Common and normal events among the elderly. *Journal of Clinical Geropsychiatry*. 4: 65-74.
Gross, R. D. 1992. *Psychology—The Science of Mind and Behaviour*. London: Hodder & Stoughton.
Guthrie, R. D. 2005. *The Nature of Paleolithic Art*. Chicago: University of Chicago Press.
Hahn, J. 1986. *Kraft und Aggression: Die Botschaft der Eiszeitkunst in Aurignacien?* Tübingen: Archaeologica Venatoria.
Halverson, J. 1991. Paleolithic Art and Cognition. *The Journal of Psychology*. 126 (3): 221-236.
Harvey, L. O. 1986. Visual memory: what is remembered? In, F. Klix and H. Hagendorf (Eds.), *Human memory and cognitive* capabilities (pp. 173-187). The Hague: Elsevier.
Harvey, G. 2005. *Animism: Respecting the Living World*. London: Hurst & Company.
Heider, F. and Simmel, M. 1944. An experimental study of apparent behaviour. *American Journal of Psychology*. 57: 243-249.
Henderson, J. M., Falk, R., Minur, S., Dyer, F. C. and Mahedevan, S. 2001. Gaze control for face learning and recognition by humans and machines. In, T. F. Shipley. and P. J. Kellman (Eds.), *From fragments to objects. Segmentation and grouping in vision* (pp. 463-481). Amsterdam: Elsevier.
Henderson, J. M. and Hollingworth, A. 1999. High-level scene perception. *Annual Review of Psychology*. 50:243-71.
Hodgson, D. 2003. Seeing the "Unseen": Fragmented Cues and the Implicit in Palaeolithic Art. *Cambridge Archaeological Journal*. 13 (1):97-106.
—. 2006a. Tracings of the mind: the role of hallucinations, pseudohallucinations and visual imagery in Franco-Cantabrian cave art.*Anthroglobe*.
—. 2006b. Altered States of Consciousness and Palaeoart: an Alternative Neurovisual Explanation. *Cambridge Archaeological Journal*. 16 (1):27-37.
http://www.anthroglobe.info/docs/hodgsond_pseudohall_061204.html 2006
—. 2008. The Visual Dynamics of Upper Palaeolithic Art. *Cambridge Archaeological Journal*. 18 (3): 341-353.
—. 2013. The Visual Brain, Perception, and Depiction of Animals in Rock Art. *The Journal of Archaeology*. Article ID342801.
http://dx.doi.org/10.1155/2013/342801

—. 2017. Closely Observed Animals, Hunter-Gatherers and Visual Imagery in Upper Palaeolithic Art. *Evolutionary Studies in Imaginative Culture.* 1(2): 59-72.

—. Hodgson D. 2018. Tracing Upper Palaeolithic Iconography: The Strange Case of Animal Tracks. *Global Journal of Archaeology & Anthropology.* 3(4): 555620. doi: 10.19080/GJAA.2018.03.555620

Hodgson, D. and Helvenston, P. A. 2006a. The Emergence of the Representation of Animals in Palaeoart: Insights from evolution and the cognitive, limbic and visual systems of the human brain. *Rock Art Research*. 23 (1): 3-40.

—. 2006b. Further thoughts on comments by Chippindale and a reply to Taçon. *Rock Art Research* 2(2): 249-255.

Hodgson, D. and Pettitt, P. 2018. The origins of iconic depictions:a falsifiable model derived from the visual science of Palaeolithic cave art and world rock art. *Cambridge Archaeological Journal.* 28(4): 591-612.

Hodgson, D. and Watson, B. 2015. The visual brain and the early depiction of animals in Europe and Southeast Asia. *World Archaeology*. 47(5): 776-791.

Hoffmann, D. l., Standish, C. D., García-Diez, M., Pettitt, P. B., Milton, J. A., Zilhão, J., Alcolea-González, J. J., Cantalejo-Duarte, P., Collado, H., de Balbín, R., Lorblanchet, M., Ramos-Muñoz, J., Weniger G-Ch., and Pike, A. W. G. 2018. U-Th dating of carbonate crusts reveals Neandertal origin of Iberian cave art. *Science.* 359 (6378): 912-915.

Horowitz, M. J. 1975. Hallucinations: An information-processing approach. In, R. K. Siegel and L. J. West (Eds.), *Hallucinations-Behaviour, Experience and Theory* (pp. 163-195). New York: John Wiley and Sons.

Horowitz, M. J., Adams, J. E. and Ruthkin, B. B. 1968. Visual imagery in brain stimulation. *Archives of General Psychiatry.* 19: 469-486.

Jaspers, K. 1963. *General Psychopathology*. (Trans. by J. Hoenig and M. W. Hamilton). Manchester, UK: Manchester University Press.

Joseph, R. 1990. *Neuropsychology, Neuropsychiatry, and Behavioral Neurology.* New York: Plenum.

Kirby, K. N. and Kosslyn. S. M. 1992. Thinking Visually. In, G.W. Humphreys (Ed.), *Understanding Vision* (pp. 71-85). Cambridge, MS: Blackwell.

Kluver, H., 1926. Mescal visions and eidetic vision. *American Journal of Psychology.* 37: 502–15.

Kosslyn, S. M. 1994. *Image and Brain.* Cambridge, MS: Mit Press/Bradford.

Kosslyn S. M. and Shin L. M. 1994. Visual Mental Images in the Brain: Current Issues. In, M. J. Farah. and G. Ratcliff (Eds.), *The Neuropsychology of High Level Vision Collected Tutorial Essays* (pp. 269-296). Hillsdale, NJ: Lawrence Erlbaum Associates.

Kosslyn, S. M. and Thompson, W. L. 2003. When is early visual cortex activated during visual mental imagery? *Psychological Bulletin.* 129: 723-746.

Kraüpl Taylor, F. 1979. *Psychopathology—Its Causes and Symptoms.* Sunbury-on-Thames: Quartermaine House.

Lange, R., Houran, J., Harte, T. M. and Havens, R. A. 1996. Contextual mediation of perception in hauntings and poltergeist-like experiences. *Perceptual and Motor Skills*. 82: 755-762.

Le Guillou, Y. 2001. Dicoveries: The Pont d'Arc Venus. *International Newsletter in Rock Art*. 29: 1-5.

Lewis-Williams, D. 2002. *The Mind in the Cave.* London: Thames and Hudson.

Lorblanchet, M. 2007. The Horse in the Palaeolithic Parietal Art of the Quercy: Outline of a Stylistic Study. In, P. Pettitt., P. Bahn and S. Ripoll (Eds.), *Palaeolithic Cave Art at Creswell Crags in European Context* (pp. 207-228). Oxford: Oxford University Press.

Malotki, E. 2013. The Road to Iconicity in the Paleoart of the American West. In, K. Sachs-Hombach and J. R. J. Schirra (Eds.), *Origins of Pictures: Anthropological Discourses in Image Science* (pp.201–29). Köln: Halem.

Manford, M. and Anderman, F. 1998. Complex visual hallucinations: Clinical and neurobiological insights. *Brain.* 121:1819-1840.

Martin, A. 2007. The Representation of Object Concepts in the Brain. *Annual Review of Psychology.* 58: 25-45.

Mcfarland, R. A. and Moore, R. D. 1957. Human factors in highway safety: a review and evaluation. *New England Journal of Medicine.* 256: 792-799.

Mormann,F., Dubois, J., Kornblith,S., Milosavljevic, M., Cerf, M., Ison, M., Tsuchiya, N., Kraskov, A., Quiroga, R.Q., Adolphs, R., Fried, I. and Koch, C. 2011. A category-specific response to animals in the right human amygdala. *Nature Neuroscience.*14:1247–1249.

Needham, A. 2017. Palaeolithic Art: More than Meets the Eye? An Object Biography Approach to Engraved Stone Plaquettes from the Magdalenian Site of Montastruc, South-Central France. Phd Thesis. University of York. http://etheses.whiterose.ac.uk/19394/

Neisser, U. 1967. *Cognitive psychology.* New York: Appleton-Century-Crofts.

Pastoors, A., Lenssen-Erz, T., Ciquae, T., Kxunta, U., Thao, T., Bégouen, R., Biesele, M. and Clottes, J. 2015. Tracking in caves: experience based reading of Pleistocene human footprints in French caves. *Cambridge Archaeological Journal.* 25(3): 551-64.

Penfield, W. and Perot, P. 1963. The brain's record of auditory and visual experience. *Brain.* 86:596-696.

Perky, C. W. 1910. An experimental study of imagination. *American Journal of Psychology.* 21:422-452.

Pettitt. P. 2002. When Burial Begins. *British Archaeology.* (August): 9-12.

Pettitt, P., Arias, P., García-Diez, M., Hoffmann, D., Castillejo, A.M., Ontañon-Peredo, R., Pike, A. and Zilhão. J. 2015. Are hand stencils in European cave art older than we think? An evaluation of the existing data and their potential implications. In, P. Bueno-Ramírez and P. G. Bahn (Eds.), *Prehistoric Art as Prehistoric Culture: Studies in Honour of Professor Rodrigo de Balbín-Behrmann* (pp.31-43). Oxford: Archaeopress.

Pettitt, P., Castillejo, A.M., Arias, P., Ontañon Peredo, R. and Harrison, R. 2014. New views on old hands: the context of stencils in El Castillo and La Garma Caves (Cantabria, Spain). *Antiquity.* 88(339): 47-63.

Pinto, Y., van Gaal, S., de Lange, F. P., Lamme V. A. F. and Seth. A. K. 2015. Expectations accelerate entry of visual stimuli into awareness. *Journal of Vision.* 15(13):1-15.

Powers, A. R., Kelley, M. and Corlett, P. R 2016.Hallucinations as top-down effects on perception. *Biological psychiatry. Cognitive Neuroscience and Neuroimaging.* 1(5):393-400.

Reeves, A. 1982. Letter to the editors. *Vision Research.* 22: 711.

Revonsou, A. 2000. The reinterpretation of dreams: An evolutionary hypothesis of the function of dreaming. *Behavioral and Brain Sciences.* 23: 793–1121.

Rilling, J. K. and Seligman R. A. 2003. A quantitative morphometric comparative analysis of the primate temporal lobe. *Journal of Human Evolution* 42:505–33.

Roebroeks, W., Sier. M. J., Nielsen. T. K., de Loecker. D., Parés, J. M., Arps, C. E. S. and Mücher. H. J. 2012. Use of red ochre by early Neanderthals. *Proceedings of the National Academy of Sciences (USA)* 109(6):1889-94.

Sandford, R. N. 1936. The effects of abstinence from food on imaginal processes. *Journal of Psychology.* 2: 129-136.

Sauvet, G. 2004. L'art Mobilier non Classique de la Grotte Magdalenienne de Bedeilhac (Ariege). In, A.-C. Welté and E. Ladier (dir.), *Art mobilier paléolithique supérieur en Europe occidentale Actes du colloque 8.3, Congrès de l'UISPP, Liège*. 2-8 Septembre 2001. (167 - 176). Liège: ERAUL.

Schacter, D. L. 1996. *Searching for Memory: the Brain, the Mind, and the Past.* New York, NY: Basic Books.

Schmid, E. 1984. The human form and the human face in Palaeolithic art. In, H. Bandi., W. Huber., M. R. Sauter. and B. Sitter (Eds.), *La contribution de la zoologie et de l'ethologie à l'interprétation de l'art des peuples chasseurs préhistoriqùes* (pp. 349-352). Fribourg, Switzerland: Editions Universitaires Fribourg.

Segal, S. J. and Fusella, V. 1970. Influence of imagined pictures and sounds on detection of visual and auditory signals. *Journal of Experimental Psychology*. 83:458-464.

Sekuler, R. and Blake, R. 1994. *Perception.* New York: McGraw Hill.

Shepard, R. N. 1984. Ecological Constraints on Internal Representation: Resonant Kinematics of Perceiving, Imagining, Thinking and Dreaming. *Psychological Review.* 91 (4): 417-447.

Straus, L. G., Morales, M. R. G. and Cuenca-Solana, D. 2015. The Magdalenian human burial of El Mirón Cave (Ramales de la Victoria, Cantabria, Spain): introduction, background, discovery and context. *Journal of Archaeological Science.* 60:1-9.

Suzuki, K., Roseboom, W., Schwartzman, D.J., and Seth, A. N. 2017. A Deep-Dream Virtual Reality Platform for Studying Altered Perceptual Phenomenology. *Scientific Reports.* 7 (15982). doi:10.1038/s41598-017-16316-2

Tulving, E. 1983. *Elements of episodic memory*. Oxford: Clarendon Press.

Turner. M.A. 2014. A Short Note on Pseudohallucinations. *Psychopathology*. 47 (4): 270-273.

Valle, R., López, G-T. E., Tuyuka, P. H.T. and Munduruku, J. S. 2018. What is Anthropogenic? On The Cultural Aetiology of Geo-Situated Visual Imagery in Indigenous Amazonia. *Rock Art Research.* 35 (2):123-144.

Vernon. M. D. 1971. *The Psychology of Perception*. Harmondsworth: Penguin.

Walker, J.W.P., Clinnick, D.T.G. and Pedersen, J.B.W. 2016. Profiled hands in Palaeolithic art: the first universally recognised symbol of the human form. *World Art*. 8(1): 1-19.

Watson, B. 2011. The eyes have it: human perception and anthropomorphic faces in world rock art. *Antiquity.* 85: 87–98.

White, R. 2003. *Prehistoric Art - the symbolic journey of humankind*. New York: Harry N. Abrams.

White, R., Bourrillon, R., Mensan, R., Clark, A. and Chiotti, L. 2017. Newly discovered Aurignacian engraved blocks from Abri Cellier: History, context and dating. *Quaternary International.* http://dx.doi.org/10.1016/j. quaint.2017.02.0015.

Whitehouse, D. 2002. Faces from the Ice Age. *BBC News: Science/ Nature.* http://news.bbc.co.uk/1/hi/sci/tech/2012385.stm

Whitlock, F.A. 1987. Pseudohallucination. In, R.L. Gregory (Ed.), The Oxford Companion to the Mind (p. 648). Oxford: Oxford University Press.

Wu. J., Duan, H., Tian, X., Wang, P. and Zhang, K. 2012. The effects of visual imagery on face identification: an ERP study. *Frontiers in Human Neuroscience*. 6:305. doi.org/10.3389/fnhum.2012.00305

Yang, J.J., Francis, N., Bellgowan, P.S.F., and Martin, A. 2005. Object concepts and the human amygdala: enhanced activity for identifying animals independent of in-put modality and stimulus format. *Paper presented at 12th Annual Meeting of the Cognitive Neuroscience Society.* New York.

Yarbus, A. L. 1967. *Eye movements and vision.* New York: Plenum Press.

Zhu, W., Drewes, J., Peatfield, N.A. and Melcher, D. 2016. Differential Visual Processing of Animal Images, with and without Conscious Awareness. *Frontiers in Human Neuroscience.* 10(513): 1-19.

# CHAPTER SEVEN

# DATES, RITUAL, AND ADAPTATION

> Unfinished animals are in fact common in Palaeolithic art, and in many cases are not a result of partial destruction of the representations (Ucko and Rosenfeld, 1967, p. 59)

## Style and chronology

Though the apparent naturalism of Chauvet appears somewhat abruptly and early that, in many respects, can be accounted for by the concept of hyperimages, a general tendency towards basic to more complex representations can still be said to take place from earlier to later periods e.g., from the Aurignacian to Magdalenian. For example, in the Cantabrian region, animal depictions show a development towards greater sophistication as the following statement confirms:

> Although one cannot be more precise, the pre-Magdalenian cycle figures do at least have a common stylistic and formal foundation. They are simple figures, represented only by the outline; usually do not depict any anatomical details, they are out of proportion and have inarticulate connections between the body and the extremities (Ochoa and García-Diez, 2015).

In fact, the very early date for Chauvet has been questioned (Jouve 2017; Pettitt and Bahn, 2015) on the basis of the authenticity of carbon dating as well as stylistic criteria. This has led some researchers to suggest Chauvet may well be Gravettian rather than Aurignacian. It seems that further dating is required to confirm the competing claims. Similarly with the sophisticated animal sculptures from the Swabian Jura, Germany that *appear* to be earlier than many of the depictions of animals from France and Spain. In this regard, doubts have been raised concerning the date of the famous Hohlenstein-Stadel "lion man" as well as related animal figurines, which may not be Aurignacian but of more recent UP vintage (Wehrberger, 2013; Jouve, 2017).

It also needs to be repeated here that the depiction of hands occurred well before Chauvet in both Europe and Indonesia (Borneo and Sulawesi). Recall that hand stencils and geometric marks produced by Neanderthals date to at least 64,000 years (Hoffman et al. 2018;Hodgson and Pettitt, 2018). In addition, a number of the earliest animal depictions, similar to those from various parts of the world (Hodgson and Watson, 2015), are now confirmed for Apollo 11 cave in southern Africa that represent a zebra, rhinoceros, and a possible therianthrope dating to around 30,000 years ago (Rifkin et al. 2015). The similarity in the way animals are depicted from these various sites derives not just from a hunter-gatherer lifestyle but, as already outlined, from the way the visual brain functions.

The full range of abbreviated to more complete cues can be seen at Gargas cave (Barrière,1976; Foucher et al. 2007) beginning around 27,000 BP. Thus, the first depictions of animals at Gargas consist of the outline of the head, neck and withers, followed by the inclusion of the forward areas and hindquarters and, finally, the complete outline. Some later animal depictions in caves dating to the Magdalenian—where fully fledged and partial outlines of animals exist—reveal a similar sequence in the drawing of animal contours to those from much earlier periods (Fritz, 1999; Fritz and Tosello, 2000). Similarly, at Rouffignac cave, where numerous outlines of mammoths can be found, research has similarly established that the mammoth's characteristic domed-shaped head was engraved first after which the dorsal contour was added (Van Gelder 2010).

Hence, even though early examples of relatively complete depictions of animals exist, the majority are partially conceived with a focus on the cervico-dorsal line. The fact that relatively complete examples of animals occur at the time when the first depictions arose suggests that as soon as the trick of representation was realised it was occasionally exploited as a template to produce more sophisticated depictions. Nevertheless, throughout the UP the main focus continued to be the incomplete animal outlines.

## Ritual concerns

A hunter-gatherer lifestyle is associated with highly-emotive, often unpleasant—but infrequent—imagistic rituals (Whitehouse, 2012). UP cave art is consistent with such demanding rituals which, besides being imagistic, is situated in hazardous locations with most cave sites only visited occasionally. Indeed, a number of caves are difficult to enter with some depictions located in the deepest, most perilous, reaches. The rituals practised seem to have involved various groups of different ages and sexes—as the evidence from handprints and foot impressions indicate (Pastoors et al. 2015; Sharpe and Van Gelder, 2006)—which hint at small mixed parties or individuals entering the caves possibly in a "quest" to locate projective hyperimages. This is borne out by the finding that the foot impressions on cave floors are relatively sparse with little indication of the kind of repetitive behavior that is consistent with frequent rituals. Rather, small groups entered the caves with some producing depictions whereas others simply explored the interior perhaps searching for the provocative suggestive natural features. Recent research (Pastoors et al. 2015), which recruited the tracking expertise of San hunters to interpret behaviour from the cave footprints, seems to support this scenario as outlined in the following passage:

> For not one of the tracks that had formerly been interpreted as 'ritual dance' or similar ceremonial behaviour was there any corroboration by the trackers. All footprints appear to have been generated by ordinary stride with a few exceptions of a faster pace, thus leaving no space for hypotheses regarding extraordinary behaviour. (Pastoors et al, 2015, p. 9)

Yet, cave art can still be considered as a place where some kind of ritual occurred probably related to a search for suggestive natural features and hyperimages. In short, it is more than likely that, if rituals were performed in caves, they were fairly rare, highly emotive, dangerous, and of short duration (Arias, 2009), and practised by small groups.

## Was early art adaptive?

As art arises from cultural evolution rather than being adaptive in strict evolutionary terms (Hodgson and Verpooten, 2014; Hodgson and Watson, 2015), the relationship between the animal depictions and evolutionary factors warrants further consideration. In light of the close affiliation between the visual brain's natural ability to detect fauna and the animal depictions, as highlighted, representing animals will have inadvertently helped sharpen the perceptual system's propensity to detect animals, thereby facilitating rapid and efficient identification of fauna on hunting forays. This has been confirmed by the fact that, when artists engage in depicting certain objects over a prolonged period, particular areas of the brain become attuned to perceiving the same objects as a result of "visual fluency" (Reber, 2012). In other words, the visual system's ability to detect animals would have been facilitated. In one sense, the depiction of animals can be viewed as adaptive but, to reiterate, only in so far as this was an unintentional byproduct. Thus, because visual imagery—as an extension of the visual brain—became an externalised graphic commodity, it could, henceforth, influence the minds of others as well as the associated cultural lifeways. For reasons already stated, the depictions ultimately became a focus of ritual/animistic practices.

In short, long-term evolutionary influences were reinforced by the everyday interaction with animals that meant it was inevitable fauna came to be depicted in the way described. The direct neural pathway from the visual cortex to the limbic/emotional brain for detecting animals also suggests animals predominate in rock art precisely because they induce a strong emotional response related to survival. It is therefore legitimate to suggest that the depiction of animals was expedited by the evolutionary prerogatives of the visual brain. Again, this does not mean animal representations rely exclusively on biological components as that would imply a specific area of the visual system evolved to *depict* animals. Rather, evolutionary factors that shaped the visual brain served to constrain the way animals were depicted during the UP. In that sense, early rock art can be said to emulate the underlying neurocognitive mechanisms. By showing that the subject matter of animals, as well as the way they were depicted, are mainly determined by neurovisual/perceptual imperatives, several enduring aspects of animal depictions can be understood in a new light.

As postulated, the detection of camouflaged animals during the majority of the Pleistocene was crucial to hominin survival to the extent that fundamental visual properties necessary for their identification became "hard wired" in the brain. This is supported by the finding that humans respond to camouflage displays in a similar way to animals (Troscianko et al. 2009). Moreover, as pointed out earlier, due to the human brain benefitting from a dedicated neural structure in the temporal cortex for the recognition of animals, certain behavioural correlates follow. For example, alterations to animals (and people) are identified in pictures more rapidly and accurately than to plants or artificial objects (New et al. 2007), and animals are automatically monitored more acutely for changes to their state and location compared to other objects.

We established earlier that imagining an object is similarly useful when the prevailing sensory stimulus is ambiguous that helps bias perception through expectation in that visual imagery primes perception. In circumstances where environmental signals are ambiguous, the visual system therefore benefits from being readily primed to regard any camouflaged feature or ambiguous surface (such as a rock face) with the slightest suggestion of fauna as an animal, which is consistent with the human visual system tendency to complete partial figures.

## References

Arias C. P. 2009. Rites in the dark? An evaluation of the current evidence for ritual areas at Magdalenian cave sites. *World Archaeology.* 41: 262–294.

Barrière, C. 1976. L'art pariétal de la Grotte de Gargas. (with the collaboration of Ali Sahly and students of the l'Institut d'art préhistorique de Toulouse). Translated into English by W.A. Drapkin. Bilingual edition. Toulouse: Mémoires de l'Institut d'Art Préhistorique No. 3. Oxford: British Archaeological Reports, Supplementary Series 14 (ii): Institute of Prehistoric Art of Toulouse.

Foucher, P., Juan-Foucher. C. S. and Rumeau. Y. 2007. *La grotte de Gargas. Un siècle de découvertes.* Édition Communautés de Communes du Canton de Saint-Laurent-de-Neste.

Fritz, C. 1999. Towards the reconstruction of Magdalenian artistic techniques: the contribution of microscopic analysis of mobiliary art. *Cambridge Archaeological Journal.* 9(2): 189–208.

Fritz, C. and Tosello, G. 2000. Observations on the techniques in the horse panel at the Chauvet Cave (Ardèche). Example of the rhinoceroses confronting each other. *International Newsletter on Rock Art.* 26:23–29.

Hodgson, D. and Pettitt, P. 2018. The origins of iconic depictions:a falsifiable model derived from the visual science of Palaeolithic cave art and world rock art. *Cambridge Archaeological Journal.* 4: 591-612.

Hodgson, D. and Verpooten, J. 2014. The Evolutionary Significance of the Arts: Exploring the Byproduct Hypothesis in the Context of Ritual, Precursors, and Cultural Evolution. *Biological Theory.*10:73-85.

Hodgson, D. and Watson, B. 2015. The visual brain and the early depiction of animals in Europe and Southeast Asia. *World Archaeology.* 47(5): 776-791.

Hoffmann, D. l., Standish, C. D., García-Diez, M., Pettitt, P. B., Milton, J. A., Zilhão, J., Alcolea-González, J. J., Cantalejo-Duarte, P., Collado, H., de Balbín, R., Lorblanchet, M., Ramos-Muñoz, J., Weniger, G-Ch. and Pike, A. W. G. 2018. U-Th dating of carbonate crusts reveals Neandertal origin of Iberian cave art. *Science.* 359 (6378): 912-915.

Jouve, G 2017. *Les problèmes de datation dans la grotte Chauvet et quelques grottes du Jura souabe.* Paris:BOD

New, J., Cosmides, L. and Tooby, J. 2007. Category-specific attention for animals reflects ancestral priorities, not expertise. *Proceedings of the National Association of Sciences USA.* 104 (42): 16598-16603.

Ochoa, B. and García-Diez, M. 2015. Chronology of western Pyrenean Paleolithic cave art: A critical examination. *Quaternary International.* 364: 272-282.

Pastoors, A., Lenssen-Erz, T., Ciqae, T., Kxunta, U., Thao, T., Bégouën, R., Biesele, M. and Clottes, J. 2015. Tracking in Caves: Experience based reading of Pleistocene human footprints in French caves. *Cambridge Archaeological Journal.* 25(3): 551-564.

Pettitt, P. and Bahn, P. 2015. An alternative chronology for the art of Chauvet cave. *Antiquity.* 89: 542-553.

Reber, R. 2012. Perceptual fluency, aesthetic pleasure, and culturally shared taste. In, A.P. Shimamura and S.E. Palmer (Eds.), *Aesthetic Science: Connecting Minds, Brains and Experience* (pp. 223–249). New York, NY: Oxford University Press,

Rifkin, R. F., Henshilwood, C. S. and Haaland, M. M. 2015. Pleistocene Figurative Art Mobilier from Apollo 11 Cave, Karas Region, Southern Namibia. *South African Archaeological Bulletin.* 70 (201):pp.113-123

Sharpe, K. and Van Gelder. L. 2006. The Study of Finger Flutings. *Cambridge Archaeological Journal.* 16: 281-295.

Troscianko, T., Benton, C. P., Lovell, P. G., Tolhurst, D. J. and Pizlom, Z. 2009. Camouflage and visual perception. *Philosophical Transactions of the Royal Society B.* 364(1516): 449–461.

Ucko, P. and Rosenfeld, A. 1967. *Palaeolithic cave art*. London: Weidenfeld and Nicolson.

Wehrberger, K. 2013. *The Return of the Lion Man: History, Myth, Magic.* Ulm: Thorbecke.

Whitehouse, H. 2012. Ritual, Cognition, and Evolution. In, R. Sun (Ed.), *Grounding the Social Sciences in the Cognitive Sciences.* Cambridge (pp. 265-284).Cambridge, MA: MIT Press.

# CHAPTER EIGHT

# CHANGING PERSPECTIVES

> Therefore, one way to naturalize aesthetics is to argue that such a system evolved first for the appraisal of objects of survival advantage, such as food sources, and was later co-opted in humans for the experience of artworks for the satisfaction of social needs. (Brown et al 2011, p. 250).

## From Upper Palaeolithic naturalism to Neolithic formalism

The explanation outlined above for the appearance of UP art centres on the role animals played in the daily lives of palaeo-humans both on an evolutionary timescale and as an ongoing activity. This led to the conclusion that image making, rather than being an activity detached from daily preoccupations, was, in fact, closely bound to activities associated with animals. It should be re-emphasised here that, although animals were a constant in the daily lives of palaeo-humans, this derived from a range of activities including hunting, warding off carnivores, consumption of meat, exploitation of hide for the making of clothes and coverings, the production of tools from bones, antlers, tusks, use of fat for lighting, construction of shelters, as well as a myriad of other purposes. In addition, the migration of hunting groups annually was dictated by the seasonal movement of fauna. Because of this dependency, animals became an overriding concern in UP art.

Besides the exploitation of natural features, a further clue to derivation of UP depictions is the pervasiveness of the typical viewing profile of animals, which you may recall, exemplifies how visual imagery/memory conserves information about animals. As the profile view provides a proficient means of storing information in memory, it thus became the primary way animals were depicted. However, from about 12,000 year ago, depictions of animals became more abstract and generalised, as in Spanish Levantine rock art (Nash and Chippindale, 2004). Here, rather than just single profiles of animals found in UP depictions, obvious scenes depict the chasing down of animals by hunters in scenes (see Figures 8-1 and 8-2). Remarkably, a similar progression can be found in Lubang Jeriji

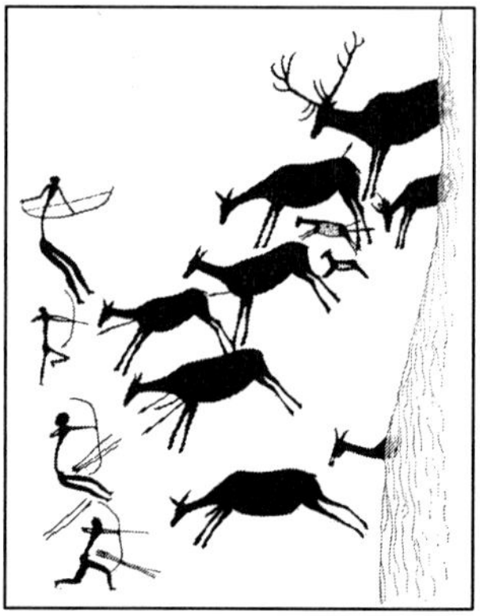

Figure 8-1. Deer hunting scene of horses at Valltorta Barranco, Eastern Spain (Obermaier and Wernert, 1919 [Public Domain]).

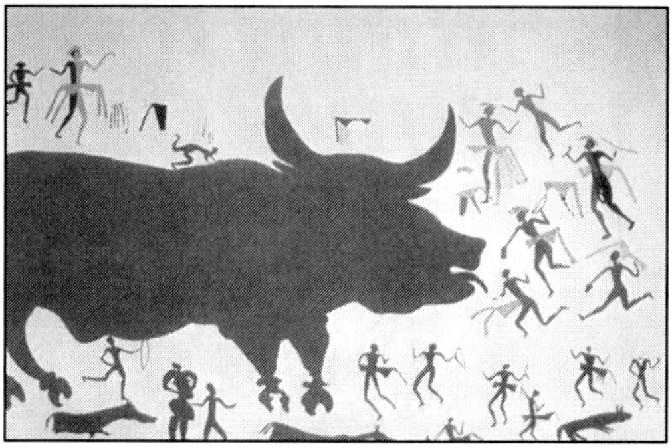

Figure 8-2. Early Neolithic copy of painting from Çatalhöyük showing hunters taunting a large herbivore. (Image by Omar Hoftun. Creative Commons Attribution-Share Alike 3.0 Unported licence).

Saléh cave in, Borneo—mentioned earlier—where the depiction of handmarks and animals is followed by scenes of humans, represented as stick-like figures hunting animals (Aubert et al. 2018). This change may reflect the fact that, during the UP, animals and humans were on an equal footing where humans, as hunters of herbivores, were at the same time targeted by carnivores. From about 30,000 years ago, carnivores and large dangerous herbivores were becoming extinct, probably due to changing environmental and climatic conditions as well as human interposition, which is reflected in a reduction in their representation during the late UP. The Mesolithic heralded a further, but decisive shift, with humans gaining ascendancy leading to the almost total demise of large carnivores, making the exploitation of the remaining large herbivores less taxing. The scenes represented no longer depict a reverence or admiration for hunted animals rather there is a gleeful running down of large fauna with weapons openly brandished (see Figure 8-1 and 8-2). The message seems to be, "we are now the masters and you, the animals, are henceforward the pursued." This is echoed by groups of hunters now depicted in scenes, with many hunters outnumbering the fauna, all skillfully captured with movement and agility. Conversely, in UP art the emphasis was always on animals with a few poorly drawn humans represented. Moreover, during the Mesolithic, representation became a means of conveying a story with animals and humans performing distinct roles. Gone is the ambiguity of superimpositions and abbreviated animal outlines as well as the apparent naturalism, which indicates hyperimages became less influential and cultural factors began to dominate. At the same time, the depictions became increasingly abstract and therefore easier to produce—a tendency that continued throughout the Neolithic.

## Changing role of "art"

The depictive strategies of the Mesolithic and early Neolithic heralded a different and developing relationship between image making, animals, and social complexity. And much time and effort was expended by increasingly settled communities towards producing new forms of art. The question arises as to why so much effort and resources continued to be allocated to "non-functional" behaviour when humans had come to dominate animals and live in protected artificial environments. In short, we need to account for the time, effort, and resources invested in making art. The issue continues to be controversial as it is not always clear why the arts first evolved and persisted, though some important preliminary waymarks have been highlighted in previous chapters. Yet, from the outset, making images would have been intertwined with other "non-functional" activities, such as singing, dancing, music, and storytelling etc., probably within a ritual context. The complexity of the issue is highlighted in the following statement:

> ...cultural evolution occurs on multiple timescales and is driven by dynamics at different levels: individuals, populations of behaving individuals, technological entities and cultural ideas. (Kolodny, et al. 2018, p. 5)

As stipulated, I argue that the arts are not a biological adaptation rather they evolved as a by-product of biological traits that were expressed through culture. It seems that, as an important attribute of the arts, the figurative depictions conceived during the UP were co-opted by later cultures in order to cement group cohesion in response to increasing population densities. As this is important to understanding how figurative representations became central to many cultures, evidence now needs to be explored in support of this possibility.

## The arts as materially embodied culture

For pre-literate and hunter-gatherer groups, art objects are invariably associated with socio-cosmological concerns and, therefore, are especially resonant as they appear to flow from the hands of individuals into the world. Hence, as exograms, they have a special status as a material extension of the mind (Malafouris,2013; Hodgson, 2000). As Hodder (2006) proposes, pre-literate communities invested in materials in the sense that social relations were "objectified" so the "made-ness" of social, ritual and mythic life became clearer. Recent empirical research appears to support the crucial nature of such socially imbued objects (Tylén et al. 2016). Thus, art objects are instilled with considerable emotional potency, not least because of the time and resources expended. In this way, materiality undergoes personification via socio-cognitive norms to the extent that the lifeways of a group become inextricably linked with artefacts (Gamble,1998; Gamble et al. 2011; Hodder, 2010). The arts were, therefore, ideally suited to express potency by way of materiality. In sum, the arts gave expression to the ongoing social matrix through what is termed entanglement (Hodder 2012). From this perspective the arts, by which I mean complex art involving different forms of expression, are incorporated into the social matrix of a community in multifarious ways.

Based on the above observations, we can surmise that *the arts ultimately served to bind individuals together when group size underwent expansion by providing materially-engaged normative, but costly, displays for assessing trustworthiness and identifying free riders.* In accordance with that definition Hodgson (2012a, 2017), based on Dunbar's and others' research (Dunbar, 2003; Gamble et al. 2011; Hamilton et al. 2007; Grove and Dunbar,2015; Gamble, 1998), showed that the arts permitted individuals to live in extended trust networks beyond the average natural group size of 150-300 (see also Straffon, 2016). The implication, here, is that the arts, as a cultural rather than biological phenomenon, provided a means by which anonymous individuals could live harmoniously in large-scale communities. Before moving forward on this, however, the relevance of group size needs to be addressed in greater detal.

## Natural group size and the arts

By creating artefacts imbued with social significance, which was probably first exploited when hunter gatherer bands came together at aggregation sites during the UP (Gamble et al. 2011; Hodgson, 2012a),[1] groups transcending the 150-300 limit were able to avoid the social tension that comes with living alongside unfamiliar others. That observation provides a potential explanation why complex arts and associated activities, as well as a sedentary lifestyle, did not appear until 260,000 years after the appearance of the anatomical modern human brain 300,000 years ago, (Hublin et al. 2017), which is referred to as "Renfrew's sapient paradox" (Renfrew, 2008). Before assessing this paradox, however, the archaeological record needs to be examined to assess whether a number of intriguing artefacts provide evidence for the appearance of "complex" art well before 40,000 years ago.

## "Art"over the long term

We have already mentioned three artefacts as art-like; the Makapanscat "face"pebble dating to 3 million years ago as well as the Berekhat Ram and Tan Tan "figurines" dating to around 350,000 to 250,000 BP respectively. As indicated, those objects do not conform to the aforementioned definition of the arts as they are essentially natural objects with an incidental resemblance to human features with minimum intentional intervention and can be regarded as passively conceived (Hodgson, 2013). That is, most of the surface of the objects remain untouched in that it was the projective capacity of the perceptual system that primarily transformed the objects into artefacts. To some extent, the same applies to the Neanderthal catlike "mask" from La Roche-Cotard (Marquet and Lorblanchet, 2014). The "mask" essentially remains a found, passive object as, despite the inserted bone wedge secured with pebbles placed into the natural stone to make the eyes—with some retouch around the stone's edge—the intervention is, nevertheless, somewhat superficial. Having said this, the mask is marginally more complex than the Berekhat Ram and Tan Tan figures in requiring slightly greater forethought.

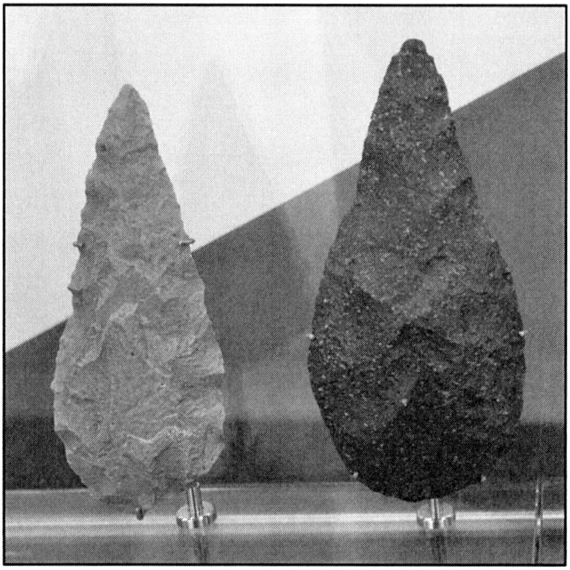

Figure 8-3. Two refined symmetrical Acheulean handaxes (Photo by the author).

Other ancient objects may also be worth considering as precursors of passive figurative artefacts. Highly symmetrical Acheulean handaxes, which began to appear in the archaeological record with greater frequency from 750,000 years ago (see Figure 8-3), display an interest in shape that goes beyond functional needs. For example, the Furze Plat handaxe from Maidenhead (c.300,000 BP), England, is far too large and heavy to use as a tool, yet displays prodigious symmetry. Such objects demonstrate a quasi/proto aesthetic concern where the interest in shape is partly decoupled from the functional (Hodgson, 2017).The beginnings of this decoupling is illustrated in a recent exhibition, where refined symmetrical Acheulean handaxes were exhibited as "first sculptures" (Berlant and Wynn, 2018) (the handaxes were displayed together with "figure stones" or "pierres figures"—natural rocks superficially suggestive of animals and human faces but which, again, can be regarded as passively conceived). Such a foundational aesthetic interest may well presage the ability to contemplate about things in the world in ways not previously entertained. What is important, at this point, is that a proto-aesthetic concern began to be expressed in material form that could potentially serve as a signal to other group members. It may therefore be that early proto-aesthetic behaviour, probably deriving from evolving cross-modal neural networks, led to a raised sense of awareness to things in the world that encouraged thinking that went beyond the everyday practicalities of survival.

Some of the brain areas implicated in aesthetic behaviour of modern humans can be informative in gauging the onset of aesthetic interest at such an early date. For example, the right inferior frontal gyrus becomes active when individuals view and produce symmetrical handaxes—an area that is related to inhibition and task shifting (Stout et al. 2015; Hecht et al 2014; Putt et al. 2017). The recruitment of that area corresponds to the same bilaterial region activated when emotionally valenced abstract artworks containing basic geometric forms are viewed (Melcher and Bacci, 2013). In effect, the enhanced interest in the symmetry of late Acheulean handaxes may signal the "coming on line" of the inferior frontal gyrus for the generalised processing of pattern and form. I direct interested readers to the footnotes (Note 1, Chapter 1) for a detailed description of the various parts of the brain implicated in aesthetic awareness.

Evidence for the "coming on line" of cerebral areas related to materially disposed proto-aesthetic behaviour consists of chevron-like motifs engraved on a shell from Java dated to 500,000 BP made by late *Homo erectus* (Joordens et al. 2015). Coloured pigment on stones, thought to be used for "artistic" purposes, has also been found in Kenya dating to a least 320,000 BP (Brooks et al. 2018). And the finding that Neanderthals made basic geometric patterns at least 64,000 years ago has led to the conclusion that the origin of non-functional mark-making is to be found in the common ancestor of *Homo sapiens* and Neanderthals (Hoffman et al. 2018).

However, only when tools began to exhibit greater variation in shape around 75,000 years ago at Blombos (Henshilwood, 2012), South Africa, does an interest in shape become *consistently* decoupled from tools. Crucially, in order to produce a greater range of tools for different specialised functions, raised levels of pattern perception is required (Hodgson 2006; Hodgson, 2016). Such an enhanced sensitivity seems to have been transferred to non-functional objects systematically at Blombos (Hodgson, 2016) where the prepared surfaces of ochre blocks were engraved with triangular shaped geometric motifs and repetitive lines (Henshilwood, 2012). Note that the engraved triangular motifs are similar in shape to the profiles of tools and repetitive straight lines simulate the tool's cutting edge. Similar engraved geometric motifs appear on ostrich eggshells around the same time (Texier et al. 2013). The South African corpus represents the first to be securely dated and accepted as part of a socio-cultural network. Still, the southern African engravings do not conform to the above definition of the arts, as most of the patterns are diminutive and were not reliable for social signalling (the same can be said of much earlier geometric marks [Hodgson, 2014]). Yet, the shell beads deriving from the same horizon may have served as necklaces and were mostly likely employed for social display (Vanhaeren et al. 2013) but within rather than between groups. This is mainly because, as body ornamentation, such items were only salient to other members of a group living in relatively small dispersed communities existing along the southern African coast at the time. Pertinently, the repetitive act of stringing shell beads not only elicits pattern perception in particular parts of the brain (Williams et al. 2015) but also cortical motor areas when repetitive actions are required to make tools (see Sherratt, 2000). Such items, however, were not a sustainable commodity due to the fact that the populations concerned were not robust or dense enough. As a result, the "art" was unable to develop into a more complex form.

## Arrival of "complex" art

It was not till after 40,000 years ago that more complex, multifaceted art appeared, as stipulated above and outlined by Conard (2008) and others (Haidle, 2014; Floss, 2018) in the form of naturalistic images, sculptures of animals, depictions of therianthropes, Venus figurines, and musical instruments such as flutes etc., (the Indonesian art from Borneo and Sulawesi is also relevant here). Some stalagmites may even have served as musical gongs (Till, 2014). Objects, such as flints and bones, were inserted into the fissures of cave walls (Clottes, 2013; Garate et al. 2013) and groups entered the caves of all ages and sexes with infants as young as two years making marks on the cave walls and ceilings (Sharpe and Van Gelder, 2006). Multiple art forms were accordingly practised by groups engaged in multi-sensory activities. This indicates that *complex* art first arose in Europe, and probably also in Indonesia and Australia, around 40,000 years ago (May et al. 2018). The question arises as to why such complexity did not appear until 260,000 years after the advent of the anatomically modern brain, which brings us back to Renfrew's sapient paradox.

## Art as a costly signalling device

Recent research indicates that population levels/densities are relevant to those issues in that, with raised numbers of individuals in a group, innovations are more likely thanks to greater opportunities to exchange information (Powell et al. 2009; Kline and Boyd, 2010; Derex et al. 2013). Also, when population levels reach a threshold, innovations are more likely to be preserved and transmitted across generations leading to cumulative effects. That scenario may be relevant to the relatively brief increase in cultural complexity in South Africa, which is associated with denser population rates along the coast (Marean, 2010). Except for critical periods due to climatic factors, in UP Europe population levels were not only rising (Atkinson et al. 2008) but did not fall below the crucial threshold for preserving and transmitting information (Shennan, 2001; Mellars and French, 2011; Powell et al. 2009). Extensive trade networks—where people frequently exchanged goods (as well as ideas) at aggregation sites (Gamble, 1998; Gamble et al. 2011)—seem to be associated with the same demographics, which probably led to the discovery/invention of figurative representations. Thus, it is pertinent that the arts became more complex and differentiated just when population levels increased and became more stable during the UP. UP communities are accordingly referred to as complex as opposed to simple hunter-gatherers (Wengrow and Graeber, 2015).

The site of Göbekli Tepe in southern Turkey sees a continuation of that tendency where a large "temple" complex acted as a meeting place in which Stone Age hunter-gatherers came together in large numbers to engage in rituals and ceremonies (Schmidt, 2010). What is revealing about this site is the amount time and effort invested in creating the various artefacts, especially as the monoliths weighed many tons. Some of these are thought to represent humans and are engraved with dangerous animals and were placed in the middle of a circular compound (Figures 8-4 and 8-5). The fact that the site was deliberately back-filled, thereby necessitating the making of similar new sites, underlines the resources directed towards the recycling of the location. For hunter-gatherers existing as relatively small roving groups, such sites seem to have facilitated group cohesion by serving as a costly signalling device at propitious times of the year.

That trend continues at Çatalhöyük in S.E.Turkey—a 10,000 year old Neolithic township thought to house a population of around 5,000—where many artefacts were seamlessly assimilated with everyday domestic activities, as can be seen in Figure 8-6 (Hodder, 2006, 2010, 2012). Despite the time and effort involved, figurines and artefacts were often deliberately destroyed or were incorporated into the walls of houses, suggesting a strong ritual tendency—perhaps in the context of ancestor worship and the reprocessing of vital forces through destruction and renewal (Fagan, 2017).

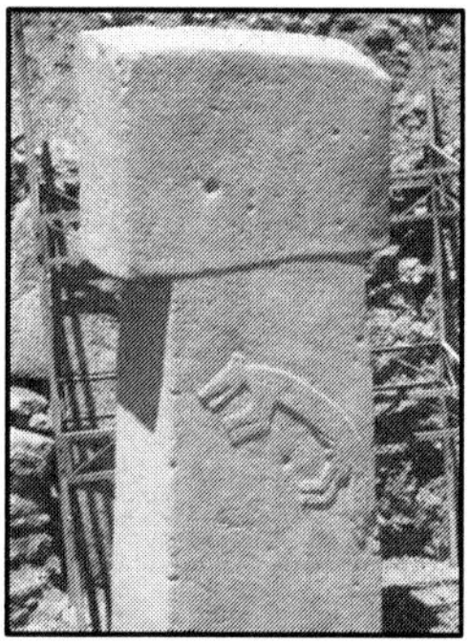

Figure 8-4. Standing stone at Göbekli Tepe. (Photo by Zhengan. Creative Commons Attribution-Share Alike 4.0 International license).

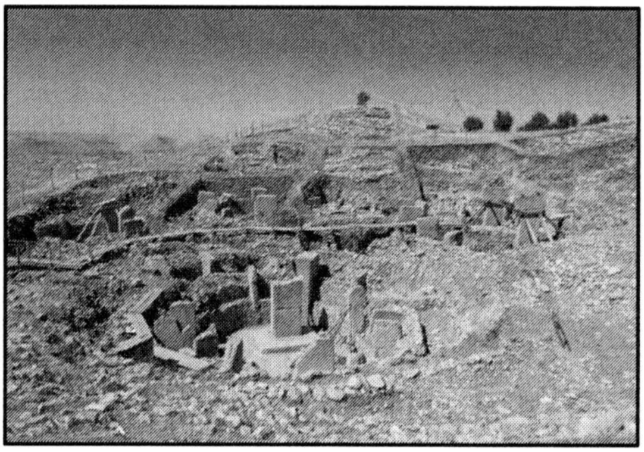

Figure 8-5. Circular compound at Göbekli Tepe with standing stones arranged in a circle. (Photo by Teomancimit. Creative Commons Attribution-Share Alike 3.0 Unported license).

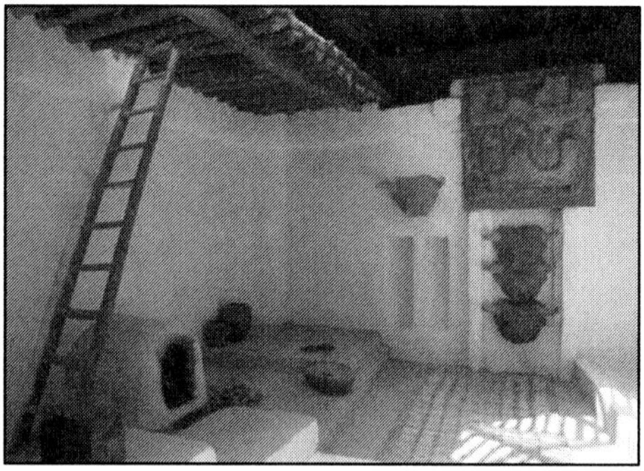

Figure 8.6. Typical room in a house at Çatalhöyük (Creative Commons Attribution-Share Alike 3.0 Unported. Author: Elelicht).

In short, the arts during the UP initially provided a way to celebrate occasions when groups temporarily assembled at aggregation sites, a trend that continued with increasing population rates as can be seen at Göbekli Tepe and Çatalhöyük. The arts provided an excellent device for that purpose as exceptional commitment to objects and activities that have no obvious survival value is required. In effect, the arts provided a means to assess the trustworthiness of others thereby allowing free-riders to be identified. In other words, before an individual could be deemed a worthy member of a group, investment in time and effort in making and participating in the arts became obligatory. Consequently, making and participating in the arts became part of the fabric of daily life at Çatalhöyük because it was necessary, in a population of around 5,000, to offset residual hostility that arises in a large resident population where many individuals may be strangers. By engaging in activities not directly related to survival, individuals were able to "prove" they were reliable members of their community. This often went along with the destruction of art objects that invariably required substantial time and effort to make; behaviour that first appeared during the UP as can be observed by the 21,000 year-old finely sculpted bison from Russia, which was deliberately broken and buried in a pit (Figure 6-6 above). Such behaviour shows that art objects, rather than existing as finished items, were part of a process of destruction and renewal probably relating to an "animistic" belief system.

In effect, the arts allowed groups to exceed the natural group size of 150-300 so that individuals could live in stable, large settled communities with many relatively anonymous individuals (Hodgson, 2012a, 2017). By making and engaging arts communally in ceremonies and rituals, a sense of wellbeing and affiliation is conferred in a way that nullifies any residual hostility towards unfamiliar others (Johnson 1982; Hamilton et al. 2007). Art objects have a special resonance because, as well as being imbued with agency through human activity, they act as a critical nexus in the diverse web of things in a culture. Material culture therefore became a vital mode of exchange and collective experience (Gosden and Marshall, 2010). Throughout the Neolithic, when large settled communities emerged, the social function of materiality not only served to cement intra group cohesion but also acted as a device for signalling to outgroups, especially as sedentary communities were increasingly obliged to protect local resources from nearby competing communities. The ability to produce iconic images provided a potent means of facilitating that process.

In short, before the UP hunter-gatherers were generally restricted to living in groups of no more than 150-300 individuals in dispersed and precarious populations. Nevertheless, during the UP when population levels increased, clans congregated in large groups beyond the 150-300 limit at aggregation sites allowing fairly complex, but loosely structured, hunter-gatherer communities to emerge at particular times of the year. For those communities, the arts became increasingly emblematic as expressed in ceremonies and rituals; a tendency which continued at Göbekli Tepe and culminated perhaps in the first settled township of Çatalhöyük. The depiction of animals and humans played a vital, perhaps pivotal role, in that process.

## Art and technology

Technological innovation and the arts are invariably viewed as unrelated activities. For hunter-gatherers and early Neolithic groups, such a dichotomy is blurred to non-existent (Van Huyssteen, 2010). Even in early settled communities, such as Çatalhöyük, a division between everyday practical activities and ritual was virtually absent (Hodder, 2010). Before the UP, we saw that a proto-aesthetic ability developed from an enhanced perceptual sensitivity for pattern and shape—allied to a specific area of the cortex—which eventually became decoupled from the visuo-spatial skills needed to make tools. During the UP, such a proto-aesthetic propensity merged with socio-cultural concerns to the extent that the arts became a collective experience that reinforced community ties.[2]

In becoming assimilated with everyday activities, the arts became a way of "playing" with materials that inevitably facilitated innovation. For example, during the Middle Stone Age in South Africa coloured ochre for body decoration was employed using heat treatment (Watts, 2002), animal figurines in UP Europe underwent hardening by applying high temperatures (Farbstein et al. 2012), and the procuring of coloured pigments for depictive purposes led to an intimate knowledge of raw materials etc., (Bahn and Vertut, 1997). At Göbekli Tepi, the first domestication of plants may have derived from the need to manage large groups during ritual events (Dietrich et al. 2012).

## Art as cultural memory device or exogram

Previously I suggested, thanks to biological factors constraining further brain expansion, material culture provided an external source of information that was able to transcend the limitations of brain size (Hodgson, 2000) in the sense it came to serve as a kind of surrogate cortex available to the community at large, referred to as an exogram (Bednarik, 2014). This supplements Donald's notion of a distributed social mind instantiated in culture (Donald, 1991) where the arts play a significant role (Donald, 2006). As part of that framework, the ability to communicate information visually through figurative depictions provided a mechanism of considerable potency that could be built upon by later generations. As stipulated, human brain size allowed groups to expand to perhaps a maximum of three hundred individuals during the Middle Palaeolithic/Stone Age. Due to the biological costs involved in further brain expansion, a natural ceiling was reached that may well have initiated the need for an external and more dependable way of dealing with information compared to speech. Depictive art provided a convenient means to meet that need. The necessity for additional expensive brain expansion was thereby avoided through the "invention" of a device for preserving information materially that also came with the advantage of an unlimited capacity. Thus, depictive art was able to make permanent what was previously intangible and temporary by providing an external information network accessible to group members (Hodgson, 2000; see also Donald, 1991, 2006); a proposal that has been taken up by Malafouris (2013) by positing the extended mind and more recently by Coward (2015) where the amalgamation of material culture into social networks allowed large scale co-operation to proceed. In sum, depictive endeavours supplied a relatively permanent means of communicating information cross-generationally as well as a way to advertise cultural identity and how a community might perceive itself.

## The place of art in evolution

The model outlined in this chapter shows that, as a cultural phenomenon, depictive art, can be considered part of the wider context of evolution. Due to its special standing, the arts are typically viewed as distinct compared to the functional/instrumental facets of material culture. As emphasised, throughout the late Pleistocene and Holocene, the arts were closely interwoven with material culture and should be considered, together with other aspects of culture, as an important means by which the extended mind was able to express prosociality. The interweaving of instrumental (technological/functional) aptitudes with conventional (prosocial behaviour) originally took place during the UP at a time when population levels were increasing, with some signs of this tentatively occurring in South Africa during the Middle Stone Age. As groups began to regularly gather at aggregation sites during the UP, raised, but temporary, population densities occurred; yet a means was required to permit anonymous individuals in such situations to interrelate in a way that appeased threat. Complex art sprang from and became essential to such gatherings, which eventually led to communities capable of living in permanent settlements with expanded populations. This does not mean that individuals with an anatomically modern brain before the UP were not endowed with the cognitive wherewithal to produce complex culture, only that the "ratchet effect" [3] (Tennie et al. 2009) was unable to operate effectively in dispersed hunter-gatherer groups with low population rates. In the final analysis, figurative depictions represent but one, yet significant, innovation that emerged out of this process.

## Notes

1. There are signs of the influence of social factors as early as 500,000 BP (Pope et al. 2006). For example, Late Acheulean handaxes are thought to have become tentatively entwined with social relationships (Coward, 2015) whereby the enhanced symmetry signaled a proto-aesthetic awareness related to social criteria (Hodgson, 2012b). As a proto-aesthetic phenomenon, however, this tendency was probably restricted to quite small isolated groups.

2. This was, to some extent, due to the fact that the arts trigger, in a disengaged way, areas of the brain that evolved to promote biological adaptation (e.g., imagination, working memory, visuo-spatial motor co-ordination) by stimulating concomitant neural domains independent of the adaptive cause thus giving rise to a sense of reassurance. In other words, the arts originated as a by-product of those cognitive domains. By auto-stimulating evolutionary instantiated brain areas through making art, the theory-of-mind network, which underpins socio-cultural affinities, was simultaneously triggered.

3. The "ratchet effect" refers to innovation that is not subsequently lost but rather is transferred to succeeding generations for improvement. The resistance to slippage or the loss of information/innovation seems to have begun in the UP and continued into the Mesolithic and Neolithic. Before the UP, innovations arose and disappeared at regular intervals probably as a result of fluctuating and unstable population levels that prevented a sustainable ratchet effect from taking hold.

## References

Aubert, M., Setiawan, P., Oktaviana, A. A., Brumm, A., Sulistyarto, P. H., Saptomo, E. W., Istiawan, B., Ma'rifat, T. A., Wahyuono, V. N., Atmoko, F. T., Zhao, J.-X., Huntley, J., Taçon, P. S. C., Howard, D. L. and Brand, H. E. A. 2018. Palaeolithic cave art in Borneo. *Nature.* 1476-4687. doi.10.1038/s41586-018-0679-9

Atkinson, Q., Gray, D.A and Drummond, A. J. 2008. mtDNA Variation Predicts Population Size in Humans and Reveals a Major Southern Asian Chapter in Human Prehistory. *Molecular Biology and Evolution,* 25 (2): 468-474.

Bahn, P. and Vertut, J. 1997. *Journey Through the Ice Age.* Berkeley: University of California Press.

Bednarik, R. G. 2014. Exograms. *Rock Art Research.* 31(1): 47-62.

Berlant, T. and Wynn, T. 2018. *First Sculpture: Handaxe to Figure Stone.* Dallas:Nasher Sculpture Centre.

Brooks, A. S., Yellen, J. E., Potts, R., Behrensmeyer, A. K., Deino, A. L., Leslie, D. E., Ambrose, S. H.,Ferguson, J. R., d'Errico, F., Zipkin, A. M., Whittaker, S., Post, J., Veatch, E. G., Foecke, K. and Clark. J. B. 2018. Long-distance stone transport and pigment use in the earliest Middle Stone Age. *Science.* 360 (6384): 90-94.

Brown, S., Gao, X., Tisdelle, L., Eickhoff, S. and Liotti, M. 2011. Naturalizing aesthetics: Brain areas for aesthetic appraisal across sensory modalities. *Neuroimage.* 58: 250-258.

Conard, N. J. 2008. A Critical View of the Evidence for a Southern African Origin of Behavioural Modernity. *South African Archaeological Society Goodwin Series.*10: 175–9.

Coward, F., 2015. Scaling up: Material culture as scaffold for the social brain. *Quaternary International*. 405 (A): 78-90.

Clottes, J. 2013. Why Did They Draw in Those Caves? *Time and Mind.* 6 (1): 7–14.

Derex, M., Beugin, M-P., Godelle, B. and Raymon, M. 2013. Experimental evidence for the influence of group size on cultural complexity. *Nature.* 503: 389–391.

Dietrich, O., Heun, M., Notroff. J., Schmidt, K. and Zarnkow, M. 2012. The role of cult and feasting in the emergence of Neolithic communities. New evidence from Göbekli Tepe, south-eastern Turkey. *Antiquity* 86: 674–69.

Donald, M. 1991. *Origins of the modern mind: three stages in the evolution of culture and cognition*. Cambridge, MS:Harvard University Press.

—.2006. The Arts and Cognitive Evolution. In, M. Turner (Ed.). *The Artful Mind* (pp. 3-20). New York: Oxford University Press.

Dunbar, R.I.M. 2003. The Social Brain: Mind, Language, and Society in Evolutionary Perspective. *Annual Review of Anthropology*. 32:163–81.

Fagan, A. 2017. Hungry architecture: spaces of consumption and predation at Göbekli Tepe. *World Archaeology.* doi:10.1080/00438243.2017.1332528

Farbstein, R., Radić, D., Brajković, D. and Miracle, P. T. 2012. First Epigravettian Ceramic Figurines from Europe (Vela Spila, Croatia). *PLoS ONE*. 7(7): e41437. doi:10.1371/journal.pone.0041437
Floss, H. 2018. Same as it ever was? The Aurignacian of the Swabian Jura and the origins of Palaeolithic art. *Quaternary International*. 491: 21-29.
Gamble, C. 1998. Palaeolithic society and the release from proximity: a network approach to intimate relations. *World Archaeology*. 29 (3): 426–449.
Gamble, C., Dunbar, R. and Gowlett. J. 2011. The Social Brain and the Shape of the Palaeolithic. *Cambridge Archaeological Journal*. 21(1): 115–35.
Garate, D., Labarge, A., Rivero, O., Normand, C. and Darricau. J. 2013. The Cave of Isturitz (West Pyrenees, France): One Century of Research in Paleolithic Parietal Art. *MDPI Arts*. 2:253-272. doi:10.3390/arts2040253
Gosden, C. and Marshall, Y. 1999. The cultural biography of objects. *World Archaeology*. 31:2: 169-178.
Grove, M. and Dunbar. R. 2015. Local objects, distant symbols: fission-fusion social systems and the evolution of human cognition. In, F. Coward, R.T. Hosfield, M. Pope and F. Wenban-Smith (Eds.), *Settlement, Society and Cognition in Human Evolution: Landscapes in Mind*. (pp.15-30). Cambridge: Cambridge University Press.
Haidle, M. N. 2014. Examining the evolution of artistic capacities: searching for mushrooms? In, C. Sütterlin, W. Schiefenhövel, C. Lehmann, J. Forster and G. Apfelauer (Eds.), *Art as behaviour. An ethological approach to visual and verbal art, music and architecture* (pp.237-251) Oldenburg: Bis-Verlag der Carl von Ossietzky Universität Oldenburg.
Hamilton, M. J., Milne, B.T., Walker, R.S., Burger, O. and Brown, J. H. 2007. The complex structure of hunter-gatherer social networks. *Proceedings of the Royal Society B*. 271: 2195–2202.
Hecht, E. E., Gutman, D. A., Khreisheh, N., Taylor, S. V., Kilner, J., Faisal, A. A., Bradley, B. A., Chaminade, T. and Stout, D. 2014. Acquisition of Paleolithic toolmaking abilities involves structural remodeling to inferior fronto parietal regions. *Brain Structure and Function*. 220(4): 2315–2331.
Henshilwood, C., 2012. Late Pleistocene Techno-traditions in Southern Africa: A Review of the Still Bay and Howiesons Poort, c.75–59k. *Journal of World Prehistory*. 25: 205–23.

Hodder, I. 2006. *Catalhöyük: The Leopards Tale. Revealing the mysteries of Turkey's ancient town.* London: Thames and Hudson.
—. 2010. Probing religion at Çatalhöyük: an interdisciplinary experiment. In, I. Hodder (Ed.), *Religion in the Emergence of Civilization: Çatalhöyük as a Case Study* (pp. 1-31). Cambridge: Cambridge University Press.
—. 2012. *Entangled: An archaeology of the relationships between humans and things*. Chichester: Wiley-Blackwell.
Hodgson, D. 2000. Art, Perception and Information Processing. *Rock Art Research.* 17(1), 3–34.
—. 2006. Understanding the Origins of Palaeoart: The Neurovisual Resonance Theory and Brain Functioning. *Paleoanthropology.* http://paleoanthro.org/journal/2006/
—. 2012a. The Consequences of Human Behaviour. *Humanities* 1: 205-228. doi:10.3390/h1030205
—. 2012b. Hominin Tool Production, Neural Integration, and the Social Brain. In, J. Cole and K. Reubens (Eds.), *Human Origins: Papers from the British Academy Lucy to Language: Archaeology of the Social Brain* (pp.41-64). (online book) https://eprints.soton.ac.uk/210797/1/HumanOriginsV1.pdf.
—. 2013. Ambiguity, Perception, and the First Representations. In, K. Sachs-Hombach and J.R.J. Schirra (Eds.), *Origins of Pictures* (pp. 372-394). (Papers from the Chemnitz Conference, Germany 2010). Köln: Halem.
—. 2014. Decoding the Blombos Engravings, Shell Beads, and Diepkloof Ostrich Eggshell Patterns. *Cambridge Archaeological Journal.* 24(1): 57 – 69.
—. 2016. Deciphering Patterns in the Archaeology of South Africa:The Neurovisual Resonance Theory. In, T. Wynn and F. Coolidge (Eds,), *Formal Models in Evolutionary Cognitive Archaeology* (pp. 133-156). New York: Oxford University Press.
—. 2017. Costly signalling, the arts, archaeology and human behaviour. *World Archaeology*. 12(1): 1-20.
Hoffmann, D. l., Standish, C. D., García-Diez, M., Pettitt, P. B., Milton, J. A., Zilhão, J., Alcolea-González, J. J., Cantalejo-Duarte, P., Collado, H., de Balbín, R., Lorblanchet, M., Ramos-Muñoz, J., Weniger, G.-Ch. and Pike, A. W. G. 2018. U-Th dating of carbonate crusts reveals Neandertal origin of Iberian cave art. *Science.* 359 (6378): 912-915.
Hublin, J-J., Ben-Ncer, A., Bailey, S. E., Freidline, S. E., Neubauer, S., Skinner, M.M. 2017. New fossils from Jebel Irhoud, Morocco and the pan-African origin of Homo sapiens. *Nature.* 546: 289–292.

Johnson G. A. 1982. Organizational structure and scalar stress. In, C. Renfrew, M, Rowlands and B. A. Seagraves (Eds.), *Theory and explanation in archaeology* (pp.389–421). New York, NY: Academic Press.

Joordens, J.C.C., d'Errico, F., Wesselingh, F. P., Munro, S., de Vos, J. Wallinga, J., et al. 2015. Homo erectus at Trinil on Java used shells for tool production and engraving. *Nature.* 518: 228–231.

Kolodny, O., Feldman, M.W. and Creanza, N. 2018. Integrative studies of cultural evolution: crossing disciplinary boundaries to produce new insights. *Philosophical Transactions of the Royal Society B.* 373(20170048):1-8.

Kline, M. A. and Boyd. R. 2010. Population size predicts technological complexity in Oceania. *Proceedings of the Royal Society B.* 277 (1693): 2559–2564.

Malafouris, L. 2013. *How Things Shape the Mind: A Theory of Material Engagement.* Cambridge, MA: MIT Press.

Marean, C.W. 2010. Pinnacle Point Cave 13B (Western Cape Province, South Africa) in context: The Cape Floral kingdom, shellfish, and modern human origins. *Journal of Human Evolution* 59: 425-433.

Marquet, J-C, and Lorblanchet, M. 2014. Les productions à caractère symbolique du site Moustérien de La Roche Cotard à Longeais (Indre-et-Loire) dans leur context géologique. *Paleo.* 25: 169-194.

May, S. K., Johnston, I. G., Taçon, P.S.C., Sanz, I.D. and Goldhah, J. 2018. Early Australian Anthropomorphs: Jabiluka's Dynamic Figure Rock Painting. *Cambridge Archaeological Journal.*28(1):67-83.

Melcher, D. and Bacci. F. 2013. Perception of Emotion in Abstract Artworks: A Multidisciplinary Approach. *Progress in Brain Research.* 204: 191-216.

Mellars, P. and French, J. C. 2011. Tenfold population increase in Western Europe at the Neandertal to modern human transition. *Science.* 333: 623–62.

Nash, G. and Chippindale, C. 2004. *The Figured Landscapes of Rock-Art: Looking at Pictures in Place.* Cambridge:Cambridge University Press.

Obermaier. H. and Wernert, P. 1919. La pinturas rupestres del Barranco de Valltorta (Castellón). Madrid: Comisión de Investigaciones Paleontológicas y Prehistóricas. Memoria, 23.

Pope, M., Russel, K. and Watson, K. 2006. Biface form and structured behaviour in the Acheulean. *Lithics.* 27:4–57.

Powell, A., Shennan, S. and Thomas. M.G. 2009. Late Pleistocene Demography and the Appearance of Modern Human Behaviour. *Science.* 324:1298–1301.

Putt, S., Wijeakumar, S., Franciscus, R.G.and Spencer, J. P. 2017. The functional brain networks that underlie Early Stone Age tool manufacture. *Nature Human Behavior.* 1(0102) doi:10.1038/s41562-017-0102

Renfrew, C. 2008. Neuroscience, evolution and the sapient paradox: the factuality of value and of the sacred. *Philosophical Transactions of the Royal Society B.* 363: 2041–204.

Schmidt, K. 2010. Göbekli Tepe—the Stone Age Sanctuaries: New results of ongoing excavations with a special focus on sculptures and high reliefs. *Documenta Praehistorica.* XXXVII: 239–256.

Sharpe, K. and Van Gelder, L. 2006. The Study of Finger Flutings. *Cambridge Archaeological Journal.* 16:281-295.

Shennan, S. 2001. Demography and Cultural Innovation: a Model and its Implications for the Emergence of Modern Human Culture. *Cambridge Archaeological Journal.* 11: 5–16.

Sherratt, 2000. Comment on Hodgson: Art, perception and (archaeological) information-processing. *Rock Art Research.* 17 (1): 29-30.

Stout, D., Hecht, E., Khreishe, N., Bradle, B. and Chaminade, T. 2015. Cognitive Demands of Lower Paleolithic Toolmaking. *PLoSONE.* 10 (4): e0121804. doi:10.1371/journal.pone.0121804

Straffon, L. M. 2016. Signalling in Style: on Cooperation, Identity and the Origins of Visual Art. In, F. Panebianco, and E. Serrelli (Eds.), *Understanding Cultural Traits: A Multidisciplinary Perspective on Cultural Traits* (pp.303-318). New York: Springer.

Tennie, C., Call, J. and Tomasello, M. 2009. Ratcheting up the ratchet: on the evolution of cumulative culture. *Philosophical Transactions of the Royal Society.* 364:2405–2415.

Texier, P.-J., Porraz, G., Parkington, J.,Rigaud, J.-P., Poggenpoel, C. and Tribolo, C. 2013. The context, form and significance of the MSA engraved ostrich eggshell collection from Diepkloof Rock Shelter, Western Cape, South Africa. *Journal of Archaeological Science* 40(9): 3412–3431.

Till, R. 2014. Sound archaeology: terminology, Palaeolithic cave art and the soundscape. *World Archaeology.* 46 (3): 292-304.

Tylén, K., Philipsen, J. S., Roepstorff, A. and Fusaroli, R. 2016. Trails of meaning construction: Symbolic artefacts engage the social brain. *Neuroimage.* 134: 105-112.

Van Gelder L. 2012. New methods and approaches in the study of finger flutings. In, J. Clottes (Ed.), *Pleistocene art of the world. Actes du Congrès IFRAO, Tarascon-sur-Ariège, septembre 2010, Symposium N° spécial de Préhistoire, Art et Sociétés.* Bulletin de la Société

Préhistorique Ariège-Pyrénées, LXVLXVI, 2010-2011, CD: pp. 1207-1220.

Vanhaeren, M., d'Errico, F., van Niekerk, K. L., Henshilwood, C. S. & Erasmus,R.M. 2013. Thinking strings: additional evidence for personal ornament use in the Middle Stone Age at Blombos Cave, South Africa. *Journal of Human Evolution.* 64(1): 500–517.

Van Huyssteen, J.W. 2010. Coding the nonvisible: epistemic limitations and understanding symbolic behaviour at Çatalhöyük. In, I. Hodder (Ed.), *Religion in the emergence of civilization: Çatalhöyük as a case study.* (pp. 99–121). Cambridge: Cambridge University Press.

Watts, I. 2002. Ochre in the Middle Stone Age of southern Africa: Ritualized display or hide preservative. *South African Archaeological Bulletin.* 57: 1-14.

Wengrow, D. and Graeber, D. 2015. Farewell to the 'childhood of man': ritual, seasonality, and the origins of inequality. *Journal of the Royal Anthropological Institute*. 21( 3): 597-69.

Williams, V.M.E., Bornman, E. and Lombard, M. 2015. Patterned bead stringing in the Middle Stone Age: a pilot neuroarchaeological study. Revised Honours Research Report, University of Johannesburg.

# CHAPTER NINE

# POSTSCRIPT

> Mathematics, art, and religion may also be construed as constrained devices, adjusted to our primate brains by millennia of cultural evolution. (Dehaene, S. 2009. p. 301).

Throughout this book the deepest reaches of the visual brain have been investigated and one factor is clear, the depictions of animals in Franco-Cantabrian art derive from a close interaction with animals and the way the visual brain functions. The preference for the portrayal of animals must therefore be seen as arising, not just from prevailing environmental/ecological conditions, but also from the long-standing evolutionary events that shaped the visual brain. Projective hyperimages also stand out as particularly influential with regard to the suggestive nature of the cave environment and rock surfaces. In sum, the explanation offered here is based on how the visual brain was shaped by entrenched evolutionary events. Such an approach offers more secure evidence than insights based on cultural interpretations not least because the latter often derive from the former. In other words, cultural criteria tend to derive from the deeper currents of human existence. It may be that culture was not the main driving force that led to UP figurative art, rather the psychological/perceptual mechanisms alluded to may be better placed to reveal its secrets. This is consistent with human experience being embodied i.e., deriving from direct experience with the world, from which culture is ultimately derived. Despite an all-embracing cultural environment, even today's artists, are still preoccupied with some of the themes integral to the earliest art; namely, a recurrent interest in animals and a proclivity to seduce the eye through manipulating figure and ground in interesting ways in order to obfuscate form (Hodgson, 2005).

The extent to which that propensity continues to fascinate today's artists can be seen in the works of contemporary artists such as Trina Merry (Figure 9-1) where deception is exploited for purely artistic purposes.

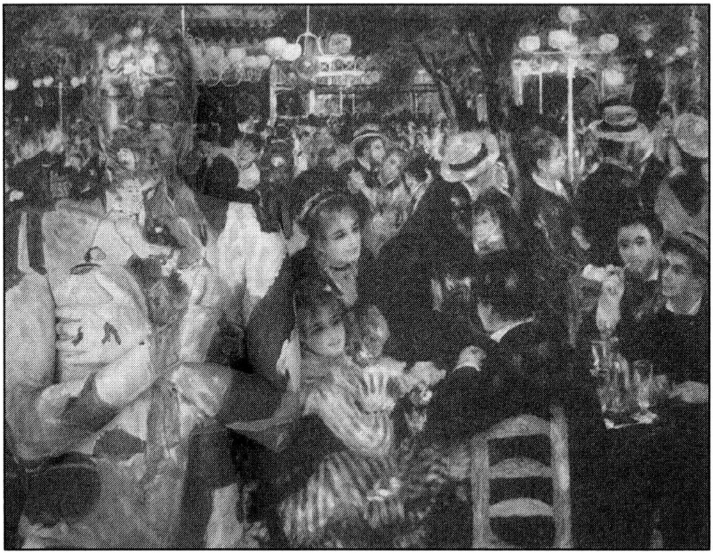

Figure 9-1. An example of intentional exploitation of camouflage created for artistic purposes by a contemporary artist, Trina Merry, where a real human figure has been painted standing in front of a Renoir painting (original in colour). (Reproduced with permission of ©Trina Merry).

What is interesting about this image is that it depicts a scene painted by Renoir in front of which stands a real person (here depicted) who blends into the actual painting. Paradoxically, the Renoir painting is just as much a visual ruse that tricks the eye into thinking something is there when it is not just as much as the camouflaged nude deceives the observer into thinking there is not something there when there is. In order to produce and unravel the nuances contained in such images, an ability to suspend disbelief is necessary, which is related to enhanced perspective taking and metacognition (Hodgson, 2013); an observation that further elucidates why fully-fledged figurative depictions took such a long time to appear.

## References

Dehaene, S. 2009. *The Reading Brain.* New York: Viking.

Hodgson, D. 2005. Graphic Primitives and the Embedded Figure in 20th Century Art: Insights from Neuroscience, Ethology and Perception. *Leonardo.* 38 (1): 55-58.

—. 2013. Ambiguity, Perception, and the First Representations. In, K. Sachs-Hombach and J.R.J. Schirra (Eds.), *Origins of Pictures* (pp. 372-394). (Papers from the Chemnitz Conference, Germany 2010). Köln: Halem.

# Index